Fields Institute Communications

VOLUME 63

The Fields Institute is a centre for research in the mathematical sciences, located in Toronto, Canada. The Institutes mission is to advance global mathematical activity in the areas of research, education and innovation. The Fields Institute is supported by the Ontario Ministry of Training, Colleges and Universities, the Natural Sciences and Engineering Research Council of Canada, and seven Principal Sponsoring Universities in Ontario (Carleton, McMaster, Ottawa, Toronto, Waterloo, Western and York), as well as by a growing list of Affiliate Universities in Canada, the U.S. and Europe, and several commercial and industrial partners.

For further volumes:
http://www.springer.com/series/10503

Panos M. Pardalos • Thomas F. Coleman
Petros Xanthopoulos
Editors

Optimization and Data Analysis in Biomedical Informatics

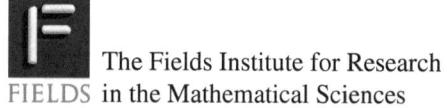

The Fields Institute for Research
in the Mathematical Sciences

Editors

Panos M. Pardalos
Center for Applied Optimization
Department of Industrial and
    Systems Engineering
University of Florida
Gainesville, FL, USA

and

Laboratory of Algorithms and Technologies
    for Networks Analysis (LATNA)
National Research University
Higher School of Economics
Moscow, Russia

Thomas F. Coleman
Department of Mathematics
University of Waterloo
Waterloo, ON, Canada

Petros Xanthopoulos
Department of Industrial Engineering
    and Management Systems
University of Central Florida
Orlando, FL, USA

ISSN 1069-5265 ISSN 2194-1564 (electronic)
ISBN 978-1-4614-4132-8 ISBN 978-1-4614-4133-5 (eBook)
DOI 10.1007/978-1-4614-4133-5
Springer New York Heidelberg Dordrecht London

Library of Congress Control Number: 2012939726

Printed on acid-free paper

Springer is part of Springer Science+Business Media (www.springer.com)

Preface

As science and society as a whole become more and more information intensive, there is an urgent need to develop, create, and apply new algorithms and methods to model, manage, and interpret this information. This is nowhere more evident than in biomedicine, where clinicians and scientists are routinely faced with conflicting (sometimes contradictory) sources of knowledge, in addition to the overwhelming and ever increasing stream of data. Bioinformatics and the -omics (genomics, proteomics, etc.) herald the advent of a new era and a new paradigm for scientific and, in particular, biomedical research. Together with the tools developed in optimization theory and the mathematical sciences, we are at a crossroads, where a more fundamental understanding of biological processes is within our grasp. This understanding will certainly pave the way for a more systematic attack on the mechanics of diseases, as opposed to a naive treatment of their symptoms (which has been the hallmark of classical medicine). It seems clear that there is an urgent need in biomedicine for new methods that will make sense out of clinical and experimental data that can be used to learn and generate rational hypotheses from the data and hence to advance the underlying disciplines.

In this volume we cover some of the topics that are related to this emerging and rapidly growing field. In June 11–12, 2010, we organizeda Workshop on Optimization and Data Analysis in Biomedical Informatics at the Fields Institute. Following this event we gathered invited contributions based on the talks presented at the workshop and additional invited chapters from world leading experts. We asked the authors to share their expertise in the form of state-of-the-art research and review chapters. Our goal was to bring together researchers from different areas and emphasize the value of mathematical methods in the areas of clinical sciences. This volume is targeted to applied mathematicians, computer scientists, industrial engineers, and clinical scientists who are interested in exploring emerging and fascinating interdisciplinary topics of research. We hope that this book will stimulate and enhance fruitful collaborations between scientists from different disciplines. The editors would like to acknowledge the Fields Institute for their financial support and hospitality. In addition, we would like to thank all the authors of the invited chapters as well as Mrs. Debbie Iscoe for her valuable help during the editing of this volume.

Gainesville, FL   Panos M. Pardalos
Waterloo, ON   Thomas F. Coleman
Orlando, FL   Petros Xanthopoulos

Contents

Novel Biclustering Methods for Re-ordering Data Matrices

Peter A. DiMaggio Jr., Ashwin Subramani, and Christodoulos A. Floudas

Abstract Clustering of large-scale data sets is an important technique that is used for analysis in a variety of fields. However, a number of these methods are based on heuristics for the identification of the best arrangement of data points. In this chapter, we present rigorous clustering methods based on the iterative optimal re-ordering of data matrices. Distinct Mixed-integer linear programming (MILP) models have been implemented to carry out clustering of dense data matrices (such as gene expression data) and sparse data matrices (such as drug discovery and toxicology). We present the capability of the optimal re-ordering methods on a wide array of data sets from systems biology, molecular discovery and toxicology.

Mathematics Subject Classification (2010): Primary 54C40, 14E20, Secondary 46E25, 20C20

The problem of data clustering is prevalent across a number of disciplines such as image processing [39], pattern recognition [3], microarray gene expression [27], information retrieval [68] and protein structure prediction [60, 74, 86]. In general, the aim of any clustering approach is to identify "similar" elements in the data set, and to organize it so that elements with similar attributes are brought together.

The most common approaches to clustering can be categorized as hierarchical [27] or partitioning [35] clustering algorithms. Although algorithms to identify the optimal solutions to these categories of problems do exist [8,71,72], most algorithms

P.A. DiMaggio Jr.
Department of Molecular Biology, Princeton University, Princeton, NJ 08540, USA
e-mail: pdimaggi@princeton.edu

A. Subramani • C.A. Floudas (✉)
Department of Chemical and Biological Engineering, Princeton
University, Princeton, NJ 08540, USA
e-mail: ashwins@princeton.edu; floudas@titan.princeton.edu

P.M. Pardalos et al. (eds.), *Optimization and Data Analysis in Biomedical Informatics*,
Fields Institute Communications 63, DOI 10.1007/978-1-4614-4133-5_1,
© Springer Science+Business Media New York 2012

end up with suboptimal solutions because of the use of heuristic search techniques and the identification of local solutions. While a number of approaches like model-based clustering [26, 84], neural networks [40], simulated annealing [44], genetic algorithms [9, 66], decomposition-based clustering [76–78], information-based clustering [73] and data classification [14, 63] have been proposed in literature, the field of rearrangement clustering has recently emerged as a very useful alternative method for minimizing the sum of pairwise distances between rows and columns to reach the optimal solution. It has been shown that this problem can be formulated as an instance of the traveling salesman problem (TSP), which can be solved to optimality [53, 54].

A bicluster is defined as a submatrix of the original matrix, which spans a subset of rows and columns. This way, common elements could be shared among a number of biclusters. This problem has been classified as an NP-hard problem [16]. An example of the application of biclustering methods is the study of downstream effects of global changes in regulated gene expression, as measured by DNA microarrays. The aforementioned clustering techniques would fail to uncover genes which are involved in more than one biological process or which are co-expressed under limited conditions [82]. This is because in an attempt to generate biclusters, most algorithms either simplify the problem representation or employ heuristic methods.

A number of biclustering algorithms have been presented in literature. The Cheng and Church algorithm [16] iteratively solves mean square residue based optimization problem using greedy heuristics. This provides a measure of the difference between the actual value of an element and its expected value based on its position in the data matrix. Since this algorithm does not transform the data, it allows for the integration of other data types. The plaid model [82] expresses data as a series of additive layers, while the spectra model [50] identifies eigenvectors which reveal the existence of checkerboard structures within the data matrix by using singular value decomposition. For a given factorization rank, the nsNMF method [15] uses non-negative matrix factorization with non-smoothness constraints to identify biclusters. The biclustering methods Bimax [65] and Samba [79] discretize the expression level which allows them to enumerate a large number of biclusters in less time than more complicated models. To complement the assortment of problem representations for biclustering, there have been a variety of algorithmic approaches developed to solve these models of varying complexity, such as zero-suppressed binary decision diagrams [85], evolutionary algorithms [10, 25], Markov chain Monte Carlo [67], bipartite graphs [79], and 0-1 fractional programming [13]. An excellent review of different bicluster definitions and biclustering algorithms can be found in [58].

One of the main applications of sparse matrix clustering is in the field of Drug discovery. Drug discovery is a tedious and expensive process, involving several phases from target identification to clinical trials [62]. One of the bottlenecks in this process is the identification of potential drug compounds, normally small organic molecules or peptides, that can achieve multiple desired biological properties [57]. Finding such lead molecules can be highly difficult even with the assistance of combinatorial chemistry and high-throughput screening [7, 38]. For example, if a single molecular scaffold has N substituent sites with S distinct functional groups that may be attached at each site, then an absolute bound on the total number of compounds in this library would be S^N (one should note that this bound is typically less in practice since bound type restrictions limit the pairing of functional groups). As S and N increases, it quickly becomes impractical to synthesize and assay all the library compounds.

A common practice is to employ quantitative structure-activity relationship (QSAR) methods [33,34,64,81] to computationally predict the biological properties of the library compounds (or at least to serve as a screening and enrichment tool to eliminate chemicals that are unlikely to have drug-like properties). All existing methods for constructing predictive QSAR models involve three basic steps [64]: (1) synthesize and assay a training set of chemicals, (2) select physical/chemical/structural descriptors that can best relate to the biological properties, and (3) construct mathematical functions that quantitatively describe and predict the biological properties by these descriptors. In practice, the reliability of these methods depends critically on the choice of suitable descriptors in step (2). It should be noted here that recently developed supervised classification methods based on mixed-integer linear programming [6, 43] have been shown to work well for the descriptor selection problem. In an unsupervised approach, the success of these methods is highly dependent upon appropriate input from the user, and hence some level of user expertise.

An adaptive substituent reordering and interpolation algorithm was proposed for estimating compound properties in combinatorial libraries from the synthesis and assaying of a small number of randomly sampled library compounds [55, 70]. Fundamentally, both QSAR and this reordering algorithm are based on the assumption that there exists an underlying physical regularity in the compound library, and this regularity can be revealed from the structure-property relationships from sampling a subset of the library compounds. However, unlike any other QSAR methods, the reordering algorithm does not require any physical/chemical/structural descriptor to operate, which makes it a very general and easy-to-apply technique for drug discovery and other molecular discovery purposes.

For a single-scaffold compound library with N substitution sites and $S_n (n = 1, 2, \ldots, N)$ substituents (functional groups) at the nth site, any compound in the library can be represented by a N-dimensional vector $\mathbf{X} = \{X_1, X_2, \ldots, X_N\}$, where X_n is the index of the substituent at the nth site and the value of X_n is an integer between 1 and S_n. As a result, a biological property y of this compound is an unknown nonlinear *descriptor-free* N-variable function

$$y = f(\mathbf{X}), \tag{1}$$

which can be approximated by other parameterized functions $g(\mathbf{X}) \approx f(\mathbf{X})$ (e.g., radial basis functions) using data collected on a small subset of M compounds in the library. $g(\mathbf{X})$ can then be utilized to estimate/interpolate the property of the unsampled compounds in the library. The reliability in approximating $f(\mathbf{X})$ by $g(\mathbf{X})$ and the subsequent predicative capability of $g(\mathbf{X})$ depend on the regularity of $f(\mathbf{X})$ and $g(\mathbf{X})$ as a function of $\mathbf{X}$, which in turn is determined by the integer assignment (i.e., the value of X_i) given to each substituent on the nth site. A key component of the algorithm is then to identify the optimal integer assignments (i.e., the optimal substituent ordering) for all substituents on all N sites so that $g(\mathbf{X})$ can correctly reveal the underlying regularity of the whole library space. This is not an easy task because the total number of possible orderings is $S_1! \times S_2! \times \ldots \times S_N!$ if each of the N sites is ordered independently. In addition, the relationship between $g(\mathbf{X})$ and the orderings can be highly complicated and not amenable to derivative-based optimization algorithms. In previous studies, the optimal orderings were identified by using search algorithms that either maximize the smoothness of $g(\mathbf{X})$ [69] or minimize the root-mean-squared difference between interpolated and actual properties of the M sampled compounds [55]. Proof-of-principle studies were performed on the laboratory data of a copolymer library [70] and a transition metal complex library [55], both of which demonstrated excellent predicative capability of the algorithm over the whole library space.

In this chapter, we introduce a new strategy for efficient substituent reordering and descriptor-free property estimation. The method views substituent reordering as a special high-dimensional rearrangement clustering problem, which eliminates the need for functional approximation and enhances computational efficiency. In comparison to functional interpolation methods, clustering techniques can be more reliable for pattern recognition in the presence of considerable data noise and therefore would be better suited for drug candidate discovery where the focus is more on identifying a subset of potential drug candidates with desired properties rather than precise quantitative property predictions. Various techniques have already been developed for the rearrangement clustering of high-dimensional data [21, 22, 53, 54, 59]. These techniques have important applications in clustering ensembles of structures from free energy calculations of oligopeptides [4, 45, 48] or proteins [46, 47], de novo protein design sequences [32, 49], and design and scheduling of batch processes [41, 56]. An important prerequisite for these existing rearrangement clustering methods is that the data matrix is dense and contains only a few missing elements. Evidently, this limitation makes these techniques inapplicable for computational drug discovery, where the goal is to make predictions on copious amounts of missing data.

The chapter is organized as follows. We present two sections describing the dense and sparse matrix clustering approaches. First, we introduce a biclustering algorithm which iteratively utilizes optimal re-ordering to cluster the rows and columns of dense data matrices in systems biology. We present several objective functions to guide the rearrangement of the data and two different mathematical models (network flow and traveling salesman problem) to perform the row and column permutations of the original data matrix. We demonstrate that this global optimization method provides a closer grouping of interrelated entities than other clustering and biclustering algorithms, produces clusters with insightful molecular functions, and can reconstruct underlying fundamental patterns in the data for several distinct sets of data matrices arising in important biological applications. Following this, we present a global pairwise similarity metric to represent the smoothness of the compound property space and introduce stochastic and heuristic search algorithms, in addition to the MILP approaches presented previously, for identifying the best substituent orderings with respect to this smoothness. Computational studies on the efficacy of the sparse matrix clustering algorithm are presented, where the proposed approach is then applied to two sparsely sampled compound libraries provided by Pfizer Inc in the *Computational Studies* section. The total number of possible compounds in these libraries is 2,418 and 14,043, respectively. Computational Studies show that the proposed methods provide excellent predictions of the library subspace that is densely populated by compounds with desired properties from sampling as low as 15% of the whole library space. Further, a synthesis strategy is then presented in the *Iterative Synthesis Strategy* section, which iterates between synthesizing a batch of compounds and using the reordering techniques to guide the synthesis of the next batch of compounds. We demonstrate that this synthesis strategy is effective in identifying lead molecules while simultaneously minimizing the total number of sampled library compounds required. Finally, the dense and sparse matrix clustering approaches are applied to Toxicology data provided during the ToxCast Data Summit, and an algorithm is presented which aims to identify a subset of in vitro assays which would be sufficient to predict the toxicity of a chemical for a given in vivo endpoint.

1   Dense Matrix Clustering

In this section, we present an overview of the components of the mathematical model for dense and sparse matrix clustering. For the clustering of dense matrices, we present two formulations, namely (1) a network flow model and (2) a traveling salesman problem (TSP) based model. We then present methods to identify cluster boundaries and iteratively bicluster submatrices of the main matrix. Following this, we present results of the application of the dense matrix clustering procedure on specific data matrix sets.

For the set of variables and equations presented in this section, the index i runs over the set of rows, while the index j runs over the set of columns. The cardinality of the number of rows and columns are represented by $|I|$ and $|J|$, respectively. We define binary variables $y_{i,i'}^{row}$ as:

$$y_{i,i'}^{row} = \begin{cases} 1, & \text{if row i is adjacent and above row i'} \\ & \text{in the final arrangement} \\ 0, & \text{otherwise} \end{cases}.$$

While the definition presented above is for rows i and i' of the matrix, equations and variables similar to the ones presented in this section can be written for columns of the matrix as well. The aim of this stage is to optimally re-arrange the set of rows and columns of a given data matrix, based on a metric of similarity provided by the user. The general expression for the objective function can be written as:

$$\sum_{i} \sum_{i'} y_{i,i'}^{row} \cdot \phi(i, i'). \tag{2}$$

Here, cost of placing two rows i and i' is given by $\phi(i, i')$, which represents the degree of similarity between two rows. The similarity between two rows can be expressed using many common expressions like relative common difference (Eq. (3)), squared difference (Eq. (4)) or root-mean squared deviation (Eq. (5)) between corresponding column elements between the rows. In addition, specialized objective functions (like *enforcing* monotonicity) can also be implemented as expressions in the objective function. The formulation of both the network flow and traveling salesman problems allows for the introduction of symmetric and asymmetric objective functions.

$$\sum_i \sum_{i'} \sum_j y_{i,i'}^{row} \cdot |a_{i,j} - a_{i',j}| \tag{3}$$

$$\sum_i \sum_{i'} \sum_j y_{i,i'}^{row} \cdot (a_{i,j} - a_{i',j})^2 \tag{4}$$

$$\sum_i \sum_{i'} y_{i,i'}^{row} \cdot \sqrt{\frac{\sum_j (a_{i,j} - a_{i',j})^2}{|J|}}. \tag{5}$$

1.1 Network Flow Model

The re-arrangement of rows and columns can be modeled as a network flow model [1, 17, 29–31, 52], where the variables $y_{i,i'}^{row}$ represent the existence of an edge between rows i and i'. Thus, the final ordering of the row permutations can be represented by an acyclic graph, with an edge connecting immediate neighbors in the final ordering of rows or columns. In addition to the variables previously presented, we introduce binary variables $y\_source_i^{row}$ and $y\_sink_i^{row}$, which indicate the row which would be the first and last rows in the network flow model, respectively. Further, continuous variables representing the flows from one node of the graph to another are represented by $f^{row}i, i'$. It should be clear that the value of variables $f^{row}(i, i)$ should be zero, since a row can never be adjacent to itself. The constraints applied to the model can be categorized as under. Further details on the mathematical expressions to represent these constraints have been presented elsewhere [21].

First, the placement of two rows adjacent to each other in the final arrangement of rows is established by the following constraints:

$$\sum_{i' \neq i} y_{i',i}^{row} + y\_source_i^{row} = 1 \qquad \forall i \tag{6}$$

$$\sum_{i' \neq i} y_{i,i'}^{row} + y\_sink_i^{row} = 1 \qquad \forall i. \tag{7}$$

This ensures that each node has exactly one neighbor above and below it in the final arrangement, unless it is a source or sink node. Next, there is exactly one source and sink node in the final re-ordered sets of nodes.

$$\sum_i y\_source_i^{row} = 1 \tag{8}$$

$$\sum_i y\_sink_i^{row} = 1. \tag{9}$$

These two sets of constraints are sufficient to ensure that all nodes in the acyclic graph have unique neighbors. However, we continue to have the possibility of cyclic orderings. In order to eliminate these, we enforce the following additional constraints.

The value of the flow entering the top node (source node) is defined to be the total number of nodes ($|I|$). Starting from this node, each subsequent node in the final arrangement will have an entering flow value $|I| - 1, |I| - 2$ and so on. The flow values ensure that we introduce an asymmetry to avoid creating circular solutions.

$$f\_source_i^{row} = |I| \cdot y\_source_i^{row} \qquad \forall i. \tag{10}$$

Flow conservation equation requires that the flow entering a node is one unit greater than that leaving the node.

$$\sum_{i'} (f_{i',i}^{row} - f_{i,i'}^{row}) + f\_source_i^{row}$$

$$- f\_sink_i^{row} = 1 \qquad \forall i. \tag{11}$$

Since we have defined the convention that $f\_source_i^{row}$ starts at $|I|$, then $f\_sink_i^{row}$ has a flow value of *zero* and thus can be eliminated from the above constraint.

Relational constraints which relate the flow variables $f_{i,i'}^{row}$ to the node variables $y_{i,i'}^{row}$, thus ensuring that flow between two nodes would be zero, unless they are immediate neighbors in the network. Finally, we assign lower and upper bounds on all flow variables. Flow between any two nodes i and i' has to be between 1 and $|I| - 1$.

$$f_{i,i'}^{row} \leq (|I| - 1) \cdot y_{i,i'}^{row} \qquad \forall (i, i') \tag{12}$$

$$f_{i,i'}^{row} \geq y_{i,i'}^{row} \qquad \forall (i, i'). \tag{13}$$

1.2 TSP Model

The main objective of the TSP Model is to find the minimum cost route to visit a list of N cities and return to the starting city. The determination of the optimal TSP path for a large number of cities continues to be a major challenge. Considering each row or column as a city in a TSP path, the re-ordering of rows and columns can be modeled as a TSP problem [18]. Just as the network flow model, rows in a matrix can be represented as nodes on the TSP path. Here, the binary variables $y_{i,i'}^{row}$ are set to 1 if row i immediately precedes row i' in the final optimal TSP path. The cost of moving from one city (i.e. row/column) to another is represented by $\phi(i, i')$. Hence, the objective function for the TSP problem can be represented as given in Eq. (2), which represents the cost incurred in "visiting" each row of the matrix exactly once.

Since the TSP problem requires that the tour start and end at the same city, we introduce a dummy row which connects the top and bottom rows in the final arrangement with edges of zero cost. The constraints applied to the model require that each city in the TSP path have exactly one neighbor before and after it in the optimal path.

$$\sum_{i'} y_{i,i'}^{row} = 1 \quad \forall i \tag{14}$$

$$\sum_{i'} y_{i',i}^{row} = 1 \quad \forall i. \tag{15}$$

In a manner similar to the network flow model, cyclic tours and subtours continue to satisfy the aforementioned set of constraints. To counter this, a number of additional constraints are implemented into TSP solvers like Concorde [5], and are beyond the scope of the discussion in this chapter. A detailed mathematical formulation of the TSP based model can be found in literature [21].

1.3 Iterative Framework

The algorithm begins by optimally re-ordering a single dimension of the data matrix. Let us denote the dimension that is re-ordered as the columns and the dimension that is not re-ordered as the rows of the data matrix. For instance, in gene expression data the columns would correspond to the time series or set of conditions over which the expression level for the genes of interest (i.e., the rows) were measured. The objective function value for each pair-wise term between neighboring columns in the final ordering is evaluated and the median of these values is computed. That is, for each column j and $j + 1$ in the *final* ordering, the median of each pairwise term of the objective function, $\phi(a_{i,j}, a_{i,j+1})$, is computed, as shown in Eq. (16).

$$MEDIAN_i \quad \phi(a_{i,j}, a_{i,j+1}). \tag{16}$$

The median was selected as the evaluating metric since it is statistically less biased to outliers than the average. Cluster boundaries are defined to lie between those columns which have the *largest* median values (since the objective function is being minimized). In other words, the median is computed for all pairs j and $j + 1$ in the final ordering and the top 10% of largest median values are selected as boundaries between the re-ordered columns. These cluster boundaries are used to partition the original matrix into several submatrices. The rows of each submatrix are then optimally re-ordered over their subset of columns and clusters in this dimension are again defined using the median value of the objective function between neighboring rows in the final ordering. The algorithmic steps for the iterative framework are presented below:

1. Optimally re-order a single dimension of the data matrix. This re-ordered dimension will be denoted as the columns.
2. Compute the median for each pair of neighboring columns in the final ordering using Eq. (16). Sort these values from highest to lowest; the largest median values define the cluster boundaries between the columns. Submatrices are defined by the columns that lie between these cluster boundaries.
3. Optimally re-order the rows of each submatrix and compute the cluster boundaries for the re-ordered rows analogous to step 2.

1.4   Determination of Cluster Boundaries

Once the optimal ordering of rows or columns is carried out, either using the network flow or TSP model, this ordering can be divided into a number of clusters for further analysis. There are two approaches that have been developed for this problem: (1) ILP based Cluster boundary determination and (2) Regression approach for identifying cluster boundaries.

1.4.1   ILP-Based Cluster Boundary Determination

We propose an integer linear programming (ILP) model to determine the cluster boundaries for a given optimal ordering. First, we identify a set of "cluster seeds" by the set *Seeds*, which consists of neighboring elements in the final ordering that are locally most similar. We also denote the set of elements that are outliers, or elements that are not cluster seeds, by the set *Outliers*. The following notation is introduced: $\bar{c}$ denotes the global average of $c(i, i + 1)$ over all i, $\sigma_{\bar{c}}$ is the corresponding standard deviation of $c(i, i+1)$ over all i, and $\hat{c}_{i,X}$ denotes the local average of $c(i', i'+1)$ for all i' within a neighborhood of $\pm X$ around element i. The sets *Seeds* and *Outliers* are constructed using the following algorithm:

1. Set *Seeds* $= \emptyset$ and *Outliers* $= \emptyset$.
2. Find the $i \notin$ *Outliers* $\bigcup$ *Seeds* with the minimum $c(i, i + 1)$ in the optimal reordering.

3. If $\hat{c}_{i,X} \leq \bar{c} - \sigma_{\bar{c}}$, then add i to *Seeds* and all other elements i' to *Outliers* within the range of $\pm X$ elements of i. Else add i to *Outliers*.
4. Return to step 2 and repeat until all elements i are examined.

Given the set of cluster seeds, *Seeds*, we will formulate an ILP model to assign all other elements to one of these initial clusters. We introduce binary variables z_i which are equal to 1 if the element is assigned to the cluster immediately preceding it in the final ordering, and 0 if it is assigned to the cluster immediately after it in the final ordering.

$$z_i = \begin{cases} 1, & \text{if element } i \text{ is assigned to the cluster seed immediately before it} \\ 0, & \text{if element } i \text{ is assigned to the cluster seed immediate after it} \end{cases}.$$

We define the sets *Behind*(i) and *InFront*(i) to denote the cluster seeds, represented by the index k, that are behind and in front of the element i, respectively. Finally, for every cluster k, we denote the set of elements that are fixed to belong to this cluster seed *a priori* by the set *Fixed*(k). For instance, if the first cluster seed contains the elements 2, 3, and 4, then *Fixed*$(1) = 2, 3, 4$.

The cost associated with the assignment of any element i into the cluster preceding or following it can be dissected into several terms:

1. The fixed cost associated with assigning element i to the cluster preceding it, which are the distances between element i and all elements initially belonging this cluster.

$$FixedCost1(i) = \sum_{i' \in Fixed(Behind(i))} c(i,i')z_i \qquad (17)$$

2. If element i is assigned to cluster $k \in Behind(i)$ and element $i' < i$ is assigned to the same cluster $k \in InFront(i')$, then we need to include the cost associated with placing these two elements in the same cluster.

$$VarCost1(i) = \sum_{i':InFront(i')=Behind(i)} c(i,i')(1 - z_{i'})z_i \qquad (18)$$

3. We also need to consider the contributions between element i and elements $i' < i$ if they are assigned to the same cluster k, which precedes these elements.

$$VarCost2(i) = \sum_{i':Behind(i')=Behind(i)} c(i,i')z_{i'}z_i \qquad (19)$$

4. Analogous expressions are derived for assigning elements to the clusters succeeding them in the final ordering. The fixed cost associated with assigning element i to the cluster after it is given by:

$$FixedCost2(i) = \sum_{i' \in Fixed(InFront(i))} c(i,i')(1 - z_i) \qquad (20)$$

5. Lastly, we need to include the cost associated with placing elements i and $i' > i$ in the same cluster k that is after these elements in the final ordering.

$$VarCost3(i) = \sum_{i':InFront(i')=InFront(i)} c(i,i')(1 - z_{i'})(1 - z_i) \qquad (21)$$

The objective function is then given by minimizing the summation of these individual contributions:

$$\min \sum_i FixedCost1(i) + FixedCost2(i) + VarCost1(i)$$

$$+ VarCost2(i) + VarCost3(i). \qquad (22)$$

Note that we must constrain the feasible cluster assignments to prevent the cross-assignment of elements. In other words, if element $i + 1$ is assigned to the cluster before it, then element i cannot be assigned to the cluster after it. The following constraint enforces this restriction:

$$z_i \geq z_{i+1}. \qquad (23)$$

The nonlinearity associated with bilinear terms in the objective function can be alleviated by defining the following binary variable:

$$w_{i,i'} = z_i z_{i'} \qquad (24)$$

and incorporating the following constraints [28] into the model:

$$w_{i,i'} \leq z_i \qquad (25)$$

$$w_{i,i'} \leq z_{i'} \qquad (26)$$

$$z_i + z_{i'} - 1 \leq w_{i,i'}. \qquad (27)$$

Minimizing Eq. (22) subject to constraint Eqs. (23), (25)–(27) provides the resulting cluster assignments for a given optimal ordering and set of cluster seeds (*Seeds*). The initial membership of the set *Seeds* is a function of the exclusion window X. We vary the value of X and select the one which results in the minimum total cluster error, which is the sum of the intra and inter cluster errors [76, 78].

This biclustering model will be applied in an iterative framework to analyze the dense in vitro assay data. The chemicals and assays will be optimally re-ordered, and then outlier in vitro assays, whose average distance to all other assays in the data is less than the distance to its nearest neighbor, will be identified and removed from the matrix. After removing the outliers, the chemical and assays are again optimally re-ordered and biclusters are defined using the aforementioned MILP model for determining cluster boundaries [23].

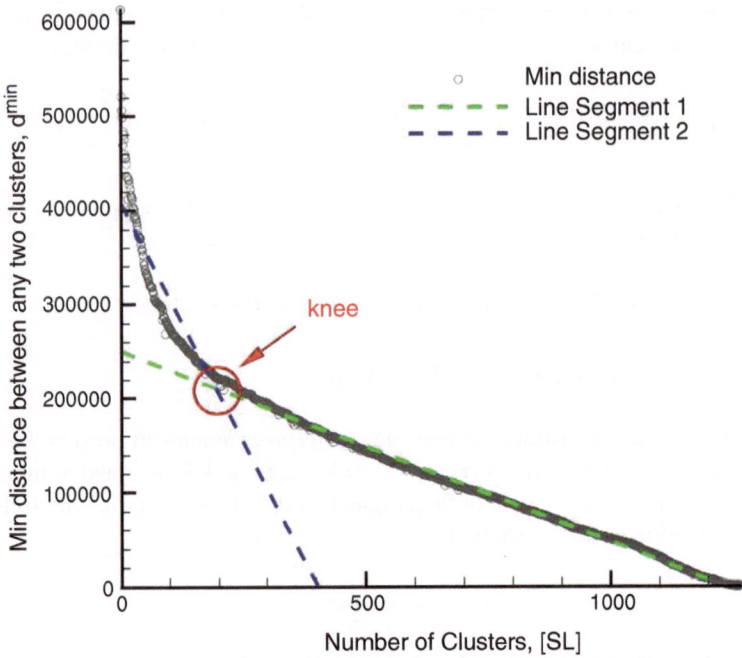

Fig. 1 Illustrative example of d^{min} as a function of SL

1.4.2 Regression Approach for Identifying Cluster Boundaries

This approach aims to identify cluster boundaries in a hierarchical manner. Again, we utilize the fact that we are aware of the optimal ordering of rows. By evaluating the distances between immediate neighbors, we can get a distribution of neighboring distances $\phi(i, i + 1)$. In this distribution, the most similar elements are merged into a cluster. Once this is done, the two elements are replaced by a single element which represents the cluster's centroid. The distance of this new "element" to the existing neighbors $i - 1$ and $i + 1$ is then evaluated and added to this distribution. At each iteration, we keep bringing elements together into clusters, and replace the elements of the cluster by the cluster centroid. It is expected that the cost of bringing elements together sequentially into clusters would keep increasing, as we keep bringing disparate elements together. As the list of elements in the distance distribution diminishes, we inevitably encounter the situation where very dissimilar elements or clusters are merged to form new clusters, resulting in a sharp increase in the cost incurred. The aim is to determine the point when the cost incurred begins to change drastically. Conceptually, this is equivalent to finding the "knee" of the curve between the cost of merging clusters and the number of elements in the neighbor list. As shown in Fig. 1, the two parts of the curve can each be approximated by best-fit line segments, and the aim would be to determine which points belong to which line segment. In order to implement the model efficiently, an Expectation Maximization (EM) based model has been developed. At each iteration, the model evaluates the probability for each point to be in one of the two classes (i.e. will be used in the next iteration to find the line of best-fit). Once these probabilities are evaluated, the points are distributed into the two classes, and each subset of points is used to re-evaluate the statistics of the best-fit line of the class to which they belong. This process is continued until convergence. A detailed description of the mathematical model and convergence strategies employed has been presented elsewhere [74].

The algorithms have been implemented in an iterative fashion. Once the optimal re-ordering and cluster boundary identification has been carried out for any dimension, the same dimensions of submatrices are subjected to further re-ordering and cluster separation. A similar procedure is employed for the other dimension of the matrix as well. The entire framework is shown in Fig. 2.

1.5   Results

For the dense matrix clustering algorithms, we present our results from six datasets, namely, metabolite concentration, image reconstruction, colon cancer, breast cancer and yeast segregant data. For each of the datasets, an overview of the data and problem size will be presented. This would be followed by a description of the clustering results [21].

1.6   Metabolite Concentration Data

The concentration profile of 68 unique metabolites has been used as the test set for this case study. The study corresponds to data from *E. Coli* and *S. cerevisiae*, observed under conditions of nitrogen and carbon starvation. The concentration values were recorded using liquid chromatography-tandem mass spectrometry [11].

The result of the application of our dense matrix clustering algorithm OREO is presented in Fig. 3. We observe that the two main cluster boundaries perfectly separate out subsets of *E. Coli* and *S. cerevisiae* conditions. Further, clustering along the columns of the matrix provided a separation between the data collected from nitrogen starvation and carbon starvation conditions. The regions between these cluster boundaries, labeled A, B, C, D, and E in Fig. 3, are also optimally re-ordered using the proposed method. In particular, for region E, re-ordering over the conditions of carbon starvation yields an excellent grouping of amino acid and TCA metabolites. In particular, 16 out of 27 metabolites in the cluster are amino acids, consistent with the observation that amino acids tend to accumulate during carbon starvation [11]. Further, the biosynthetic intermediates also order well, with all 12 being placed in the top half of the matrix.

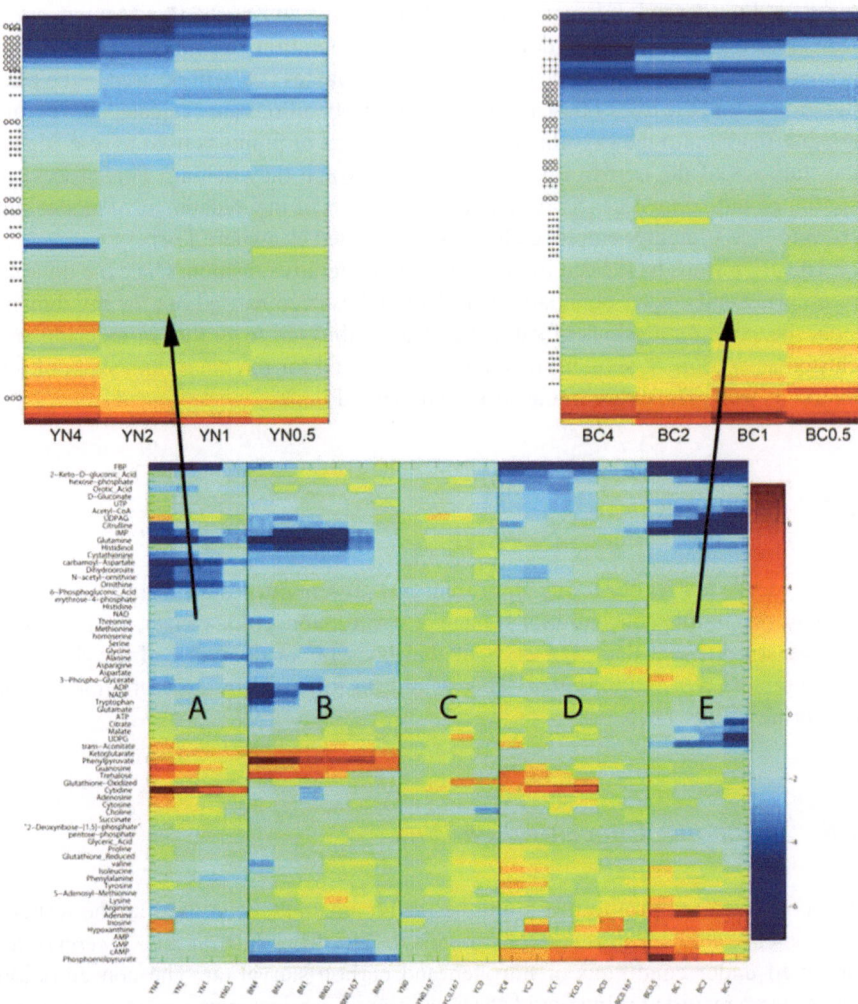

Fig. 2 Re-ordered matrix from metabolite concentration data. The two sub-matrices on the *top* reflect the improved grouping of biosynthetic intermediates and amino acids on the *left* and *right*, respectively

1.7   Image Reconstruction Data (Lenna Matrix)

In an experiment presented in [61], an image commonly referred to as the "Lenna" image of size 512 by 512 pixels, was elongated row-wise by replicating it 10 times, thus creating a matrix of size 5,120 by 512 pixels. An ideal algorithm would present a stretched version of the original image at the end of the clustering procedure.

Using our dense matrix clustering algorithm, we were able to recover the correct ordering of the original image and a subset of the original image, as shown in

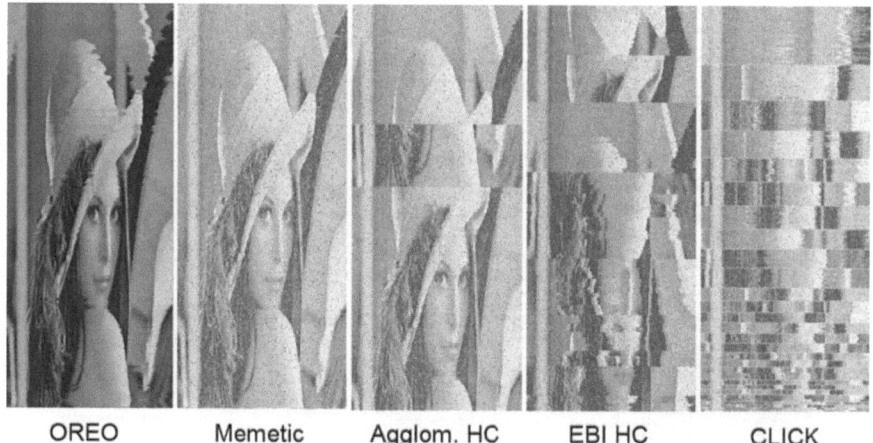

| OREO | Memetic | Agglom. HC | EBI HC | CLICK |

Fig. 3 Comparison of the re-ordering results of the Lenna matrix by different methods

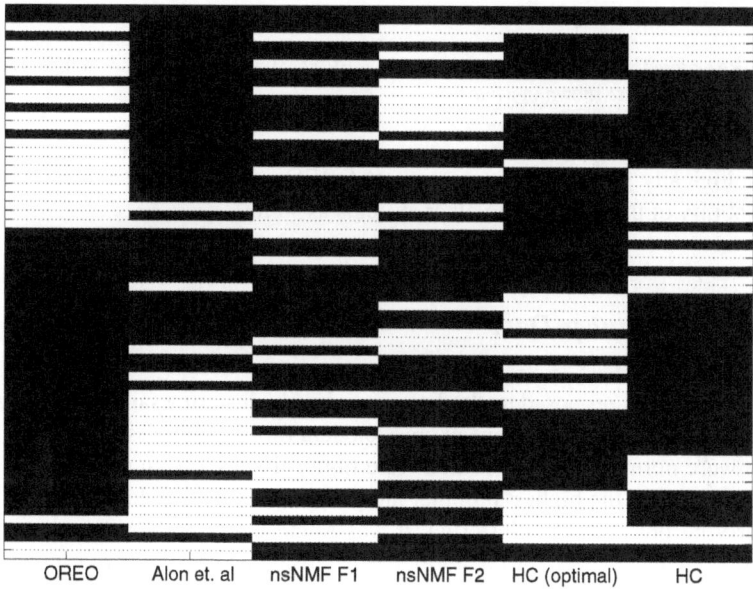

| OREO | Alon et. al | nsNMF F1 | nsNMF F2 | HC (optimal) | HC |

Fig. 4 Re-ordered matrix from colon cancer data. Tumor tissues are shown in *black*, while normal tissues are shown in *white*

Fig. 4. In addition, two kinds of noise were introduced into the data, namely (1) modifying *every* pixel by a random value less than 10% of the maximum pixel intensity (255) and (2) assigning a random value between 0 and 255 to 10% of the pixels (e.g., 262,144 of the pixels). On clustering the modified data matrix, our dense matrix clustering algorithm OREO recovers the correct image. While this dataset is

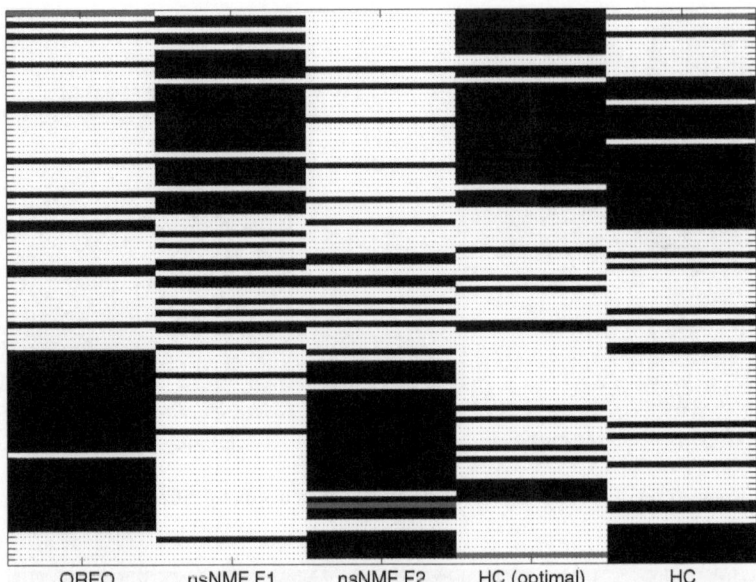

Fig. 5 Re-ordered matrix from breast cancer data. ER negative tumors are shown in *black*, while ER positive tumors are shown in *white*

not biologically generated, the success of OREO with this dataset represents the applicability of the approach to other systems as well.

1.8 Colon Cancer Data

We also tested the proposed method on a standard biclustering sample classification example [58] comprised of gene expression data for 62 colon tissue samples, 22 of which were normal and 40 of which were tumor tissues [2]. In the original work, 2,000 genes with the highest minimal intensity across all samples were examined. By carrying out two-way clustering, it was found that the algorithm was able to separate out the tissues into a normal-rich cluster and a tumor-rich cluster. By re-ordering over the set of rows (i.e. genes), OREO grouped 30 out of the 48 ESTs homologous to ribosomal proteins in one cluster, compared to 22 out of 48 that were brought together in the original work [2]. The final re-ordered matrix is shown in Fig. 5, where the tumor tissues and normal tissues are shown in black and white, respectively. Comparisons to clustering results by some other methods is also shown.

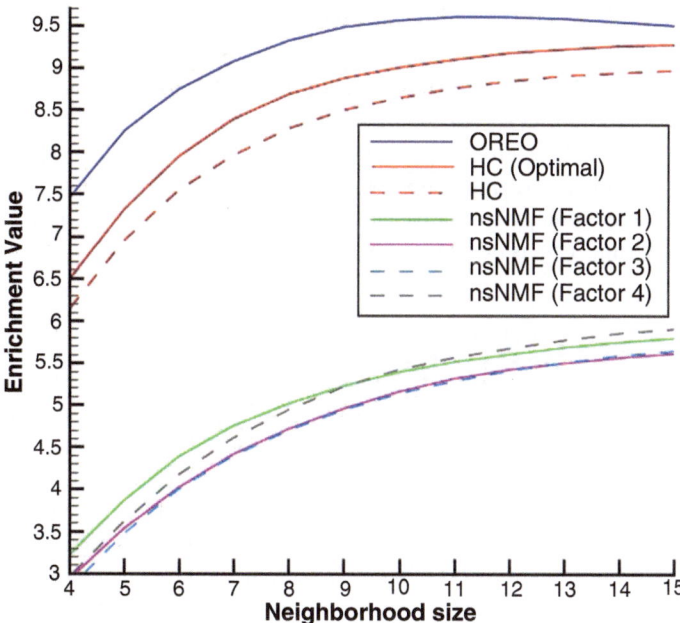

Fig. 6 Averaged enrichment values over neighborhood sizes in the range of 4 to 15 genes

1.9   Breast Cancer Data

OREO was applied to breast cancer data from literature [83], where the expression
levels for approximately 25,000 genes over 98 breast cancer tumors were measured.
In the study, a set of 5,000 genes were identified to be effective in separating ER
positive from ER negative tumor samples. These genes have at least a twofold
difference with a p-value less than 0.01 in five or more samples. The main aim of the
study was to identify a set of reporter genes for classification based on prognosis,
ER status and BRCA1 germline mutation carriers [83].

The column re-ordering results are presented in Fig. 6. As can be seen from the
figure, OREO is fairly successful in separating the ER positive (white) from the ER
negative (black) tumors. Only 13 of the ER negative tumors get assigned to the ER
positive region, while only 1 ER positive tumor gets assigned to the ER negative
region. On carrying out row-wise clustering, we observed the grouping of the 550
optimal ER status genes as determined by Van't Veer et al. [83]. Here, we found that
50 of the optimal ER status genes were found in a span of only 171 genes (29.2%),
which was a denser cluster than those observed by other clustering methods [21].

1.10 Yeast Segregant Gene Expression Data

For this dataset, the OREO methodology was applied to 6,216 genes subject to 131 stress conditions [12]. For this data, re-ordering over all columns was carried out to identify the best grouping of all genes. In order to evaluate the biological significance of the re-ordered genes, we evaluated the average enrichment for each of the 130 gene ontology terms over all possible neighborhoods of size L genes in the final ordering. This is defined mathematically as:

$$\text{Enrichment of process k} = \frac{\left(NG_L^k - 1\right)/L}{NG^k/NG} \tag{28}$$

where NG_L^k denotes the number of genes in a neighborhood of size L for process k, NG^k denotes the number of genes for process k in the entire experiment, and NG represents the total number of genes in the experiment. Equation (28) is applied for every process over all possible neighborhoods of genes, where the initial neighborhood of genes is comprised of genes of 1 though L in the final ordering and this neighborhood window is incremented by one gene (i.e., the next neighborhood contains genes 2 through L+1) until the last gene in the final ordering has been reached. The enrichment values in Eq. (28) are then averaged over the total number of neighborhoods considered. This process is repeated for several gene neighborhood sizes in the range of 4–15 genes and the results comparing our method to hierarchical clustering are shown in Fig. 7.

As can be seen from Fig. 7, OREO achieves a much improved enrichment on average than other methods. Physically, this means that genes annotated to similar biological processes are arranged closer in the final re-ordering results of OREO when compared to other methods.

The results presented demonstrate the general application of OREO to biological and non-biological data from a number of fields. Further details on implementation and run-times can be found in literature [21].

2 Sparse Matrix Clustering

The clustering of sparse matrices cannot be carried out by the methods presented in the previous section. This is because all missing elements would be considered "similar", whereas any conclusions drawn on the similarity of these elements would be erroneous. First, we define the objective function that will provide a measure of smoothness in the property space, while accounting for missing elements. This is followed by three formulations for the determination of the best ordering of rows and columns. The first method is based on deterministic optimization techniques, and can hence provide a guarantee for convergence to the optimal solution. Genetic algorithms based approaches provide solutions that are frequently very close to the

Fig. 7 Flow diagram for the iterative framework for biclustering, OREO

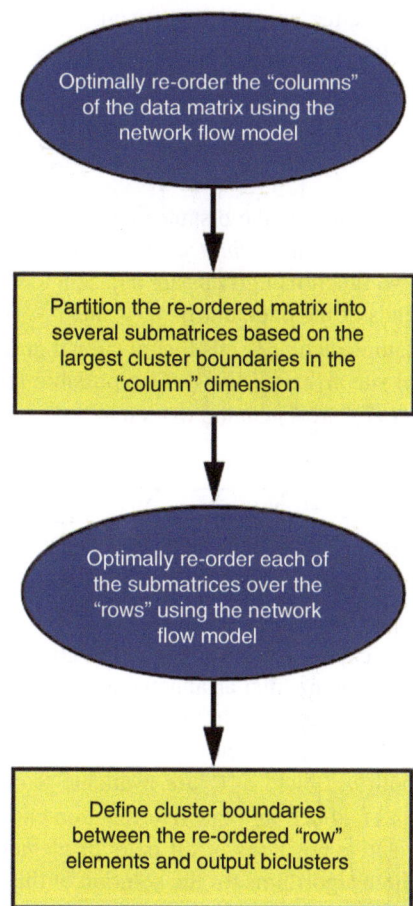

optimal solution, but lack the theoretical guarantee associated with deterministic methods. Heuristic approaches provide the potential of faster identification of good solutions to a given problem.

2.1 Objective Function

Unlike the dense matrix clustering objective function, the sparse matrix objective function cannot be restricted to comparisons between immediate neighbors. This is because, given the possibility of multiple missing values, the comparison between an element (row/column) and its neighbor would be incomplete at best. Hence, a better expression would be one which will measure the "quality" of the entire permutation of rows or columns, and not just work with immediate neighbors.

We define an objective function which depends on compound properties as well as the distance between the elements: This is given by Eq. (29)

$$Q = \sum_{m=1}^{N} \sum_{n=1,n \neq m}^{N} \sum_{j=1}^{S_n} \sum_{i=1}^{S_m} \sum_{i'=1}^{S_m} \theta(d_{i,i'}^m) \cdot \phi\left(a_{i,j}^{m,n}, a_{i',j}^{m,n}\right) \tag{29}$$

where $d_{i,i'}^m$ is the distance between functional groups i and i' at the mth site with respect to their *final positions* in the matrix. This distance value will be 1 if the two functional groups are adjacent to each other and $S_m - 1$ if they are on opposite ends of the data matrix. $a_{i,j}^{m,n}$ and $a_{i',j}^{m,n}$ denote the measured property values of two sampled compounds for functional group j in site n and functional groups i and i' in site m, respectively. One possible form of the component functions $\theta(d_{i,i'}^m)$ and $\phi(a_{i,j}^{m,n}, a_{i',j}^{m,n})$ can be written as

$$Q = \sum_{m=1}^{N} \sum_{n=1,n \neq m}^{N} \sum_{j=1}^{S_n} \sum_{i=1}^{S_m} \sum_{i'=1}^{S_m} \left(\frac{1}{w^m} \cdot \frac{S_m - d_{i,i'}^m}{S_m - 1} \right) \cdot \left(a_{i,j}^{m,n} - a_{i',j}^{m,n}\right)^2 \tag{30}$$

where $\theta(d_{i,i'}^m)$ is linear with respect to $d_{i,i'}^m$, achieving a maximum value of $1/w^m$ at $d_{i,i'}^m = 1$ and a minimum value of $1/w^m \cdot 1/(S_m - 1)$ at $d_{i,i'}^m = S_m - 1$. Thus, this expression gives the largest contributions to those elements which are grouped close in the final arrangement and a lower weight to those elements that are distant from one another in the final matrix ordering. $\phi(a_{i,j}^{m,n}, a_{i',j}^{m,n})$ is the squared property difference between two compounds. w^m is the number of compound pairs where both $a_{i,j}^{m,n}$ and $a_{i',j}^{m,n}$ are available from synthesis and property assaying for all i and i'. Such an objective function ensures that the final arrangement accounts for relative positions of all pairs of elements in the original dataset. We now present three algorithms for the solution of this problem.

2.2 Mixed-Integer Linear Optimization Based Algorithm

We define binary variables $y_{i,k}$, which indicates the assignment of element i to position k in the final ordering, such that $1 \leq k \leq |I|$, with $|I|$ being the number of elements in the dataset.

$$y_{i,k} = \begin{cases} 1, & \text{if row } i \text{ is assigned to position } k \text{ in the final ordering, 0 otherwise} \\ 0, & \text{otherwise} \end{cases}.$$

In addition, positive variables p_i are introduced, which relate the binary variables to their respective positions in the final ordering. Mathematically, this can be represented as:

$$p_i = \sum_{k} k \cdot y_{i,k} \quad \forall i. \tag{31}$$

The constraints that are applied to the model are listed below. Firstly, a final position can only contain one element, and each element can only be assigned to one final position.

$$\sum_k y_{i,k} = 1 \quad \forall i \tag{32}$$

$$\sum_i y_{i,k} = 1 \quad \forall k. \tag{33}$$

The position variables p_i can vary only between the lower limit (i.e. 1) and the upper limit (i.e. $|I|$).

$$1 \leq p_i \leq |I| \quad \forall i > 1. \tag{34}$$

The problem of symmetry can be alleviated, without loss of generality, by ensuring that the first row be placed in any one pre-determined half of the final matrix.

$$1 \leq p_1 \leq \lfloor |I| + 1/2 \rfloor. \tag{35}$$

The difference between the position variables p_i and p'_i for any pair of elements i and i' has to lie between a lower limit (i.e., 1) and an upper limit (i.e., $|I| - 1$). We represent these distances by the positive variables $d_{i,i'}$.

$$1 \leq d_{i,i'} \leq |I| - 1 \quad \forall i, i' > i. \tag{36}$$

The distance variables $d_{i,i'}$ are related to p_i and p'_i by

$$d_{i,i'} \geq p_i - p_{i'} \quad \forall i < i' \tag{37}$$

$$d_{i,i'} \geq p_{i'} - p_i \quad \forall i < i'. \tag{38}$$

While these set of constraints are sufficient to solve the problem to optimality, the resulting linear programming relaxation would not be tight enough, resulting in very long solution times. A few constraints are added to tighten the relaxations, thus improving the convergence rate to the final solution. Firstly, the summation of all final distances between elements would be a constant determined by the number of elements in the dataset.

$$\sum_i \sum_{i'>i} d_{i,i'} = C. \tag{39}$$

For instance, if there are only four rows, then $C = 3 \cdot 1 + 2 \cdot 2 + 1 \cdot 3 = 10$. Further, the summation of distances between an element i and all other elements would be a function of its final position k. For example, if row i is assigned to position 2 out of 4, then the summation of distances between it and all other elements would be pre-determined to $(2 - 1) + (3 - 2) + (4 - 2) = 4 = G(2)$. Next, the distance between any pair of rows i and i' can be written as:

$$d_{i,i'} \leq \sum_k F(k) \cdot y_{i,k} \quad \forall i < i' \tag{40}$$

$$d_{i,i'} \leq \sum_k F(k) \cdot y_{i',k} \quad \forall i < i' \tag{41}$$

where $F(k)$ is a shorthand representation of $MAX(|I| - k, k - 1)$. Since the distances are Euclidean, triangle inequality can be imposed as an additional condition.

$$d_{i,i'} \leq d_{i,i''} + d_{i'',i} \quad \forall i, i', i''. \tag{42}$$

Subsets of the aforementioned additional constraints are introduced dynamically into the model, as the entire set of additional constraints may lead to memory issues in commercial solvers for moderately sized problems. Constraints such as triangle inequality are added only to subsets of elements which violate them. This mixed-integer linear programming (MILP) formulation guarantees identification of the global optimal solution, when solved to optimality. In addition to the model presented here, we can add heuristic approaches (like the ones presented in this chapter) to identify good integer solutions, which would help convergence to the final optimal solution faster.

2.3   Genetic Algorithms Based Approach

Genetic Algorithms (GAs) are a category of popular stochastic algorithms, whose general applicability make them suitable for a wide range of problems. The solution method typically starts from a random set of element orderings. Based on the Q value for any given set of orderings, operations known as mutations and crossovers are performed to generate a modified population of element orderings. With the overall aim still being the minimization of Q values in the working population, the algorithm is continued until convergence criteria are met. The computational complexity of GAs depend on the size of the initial population. The success of the algorithm depends on the selection of a number of initial parameters, which are problem dependent. A number of approaches have been introduced which aim to integrate GAs with deterministic optimization algorithms in order to provide a theoretical guarantee to the final solution.

2.4   Heuristic-Based Approaches

A number of heuristics can be applied to either reach local minima solutions, or to support deterministic algorithms in the aim to find the global minimum. A few examples of heuristic approaches to identify integer solutions to the problem are:

1. Random swapping of rows and columns. The swapped matrix can be accepted if it has an improved Q value, or based on a probability function in a Monte Carlo based approach.

2. Greedy solutions based on selecting rows or columns to be neighbors based on minimum distance between unconnected elements
3. Ordering rows or columns based on the sum of a randomly selected subset of points along the rows and columns of the matrix

The main advantage of heuristic techniques is their scalability to large-scale problems. While they do not provide any theoretical guarantees about the final solution, empirical evidence suggests that solutions obtained by these methods are often within a few percentage points of the optimal solution. For our implementation, we integrate heuristic approaches into the mixed-integer linear programming approach, in an attempt to speed up the identification of good integer solutions. We have utilized the heuristic callback function in the CPLEX callable library [19] for the incorporation of the aforementioned heuristic based on swapping rows or columns. Further details on the implementation of the algorithms can be found in previous work by DiMaggio et al. [24].

2.5 Results

The application of the MILP model presented in the previous section to molecular discovery applications was tested on two data matrices provided by Pfizer Inc., where no information regarding the identity of compounds was available to us. The overview of the results achieved are presented as under. Further details of all results obtained are available elsewhere [24].

2.6 Data Matrix 1: Moderate-Size Compound Library

The first data matrix analyzed contains 62 rows and 39 columns of percent inhibition data for an unknown set of compounds. The most desirable compounds are the strongest inhibitors of an unknown target, which correspond to the highest percentage inhibition values in this set. The original matrix has 1,229 data values out of a possible 2,418 (51%). The optimal reordering identified by the deterministic algorithm for the full set of data values is shown in Fig. 8(f). The results for the deterministic method applied to a sampling of 30 and 50% of the available data (or 15 and 25% of the library space) are provided in Fig. 8(b) and 8(d), respectively. For predictive assessment, the placement of the un-sampled data values for the re-orderings for 15 and 25% of the full library are also revealed in Fig. 8(c) and 8(e). Contrasting Fig. 8(c) and 8(e) shows that increasing the sampling of data values from 15 to 25% of the library space improves the quality of both the overall ordering and the selected subregion, as expected. The reordered data matrices show a tendency to group the desired compounds into an easily identifiable subset of the matrix. Let us define rows 1–9 and columns 1–11 of Fig. 8(d) to be Region 1 and

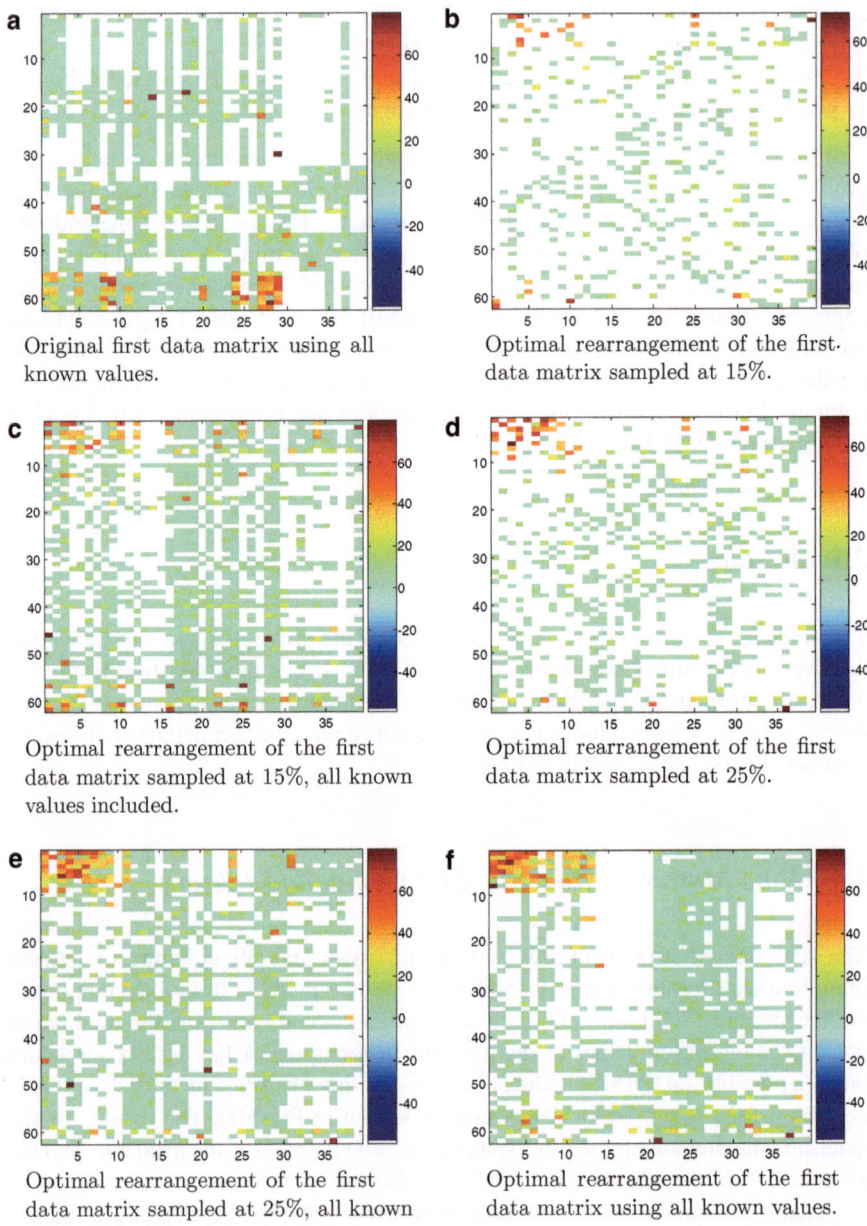

a Original first data matrix using all known values.

b Optimal rearrangement of the first data matrix sampled at 15%.

c Optimal rearrangement of the first data matrix sampled at 15%, all known values included.

d Optimal rearrangement of the first data matrix sampled at 25%.

e Optimal rearrangement of the first data matrix sampled at 25%, all known values included.

f Optimal rearrangement of the first data matrix using all known values.

Fig. 8 First data matrix reordered by deterministic optimization

the remaining matrix to be Region 2. When examining the reordered compounds for 25% of the whole library space, we see that Region 1 contains a much higher average inhibition value and a disproportionate number of data values above inhibition values of 40 and 60. Of the 12 total compounds with greater than 60% inhibition from the full data set, six sampled and three unsampled compounds are found in Region 1. The consistent abundance of high inhibition values would suggest the unused compounds in Region 1 as good candidates for future synthesis.

The stochastic optimization method and the property R sorting method also result in an aggregation of the compounds with high inhibition values for the full and sampled data matrices. The substituents found in Region 1 are very similar to those identified by the deterministic method but are less prominent in terms of the number of high inhibition compounds contained in Region 1.

2.7   Data Matrix 2: Large-Size Compound Library

The second data matrix contains 151 rows and 93 columns of percent inhibition data for an unknown set of compounds. The initial ordering of the compounds was provided by Pfizer Inc. This matrix is the largest data set that was analyzed using the proposed methods. Similarly as the first data matrix, the most desirable compounds for further study are the ones with strongest inhibition. The original matrix has 4,110 known data values out of a possible 14,043 (i.e., 29%). This data matrix was reordered using all known values and a sampled subset of known values to test the reliability of the approach. The best identified reordering of this data matrix using all of the known values (4,110 values) is presented in Fig. 9(d). The best reordering for a sampling of 50% of the available data values (or 15% of the whole library space) is presented in Fig. 9(b) and the corresponding placement of the unsampled compounds based on this reordering is revealed in Fig. 9(c). The proposed deterministic method is able to group many of the desired compounds in an easily identifiable subset of the matrix. Let us define rows 1–21 and columns 1–20 of Fig. 9(d) to be Region 1 and the remaining matrix to be Region 2. Here, it was seen that Region 1 contains a much higher average inhibition value and also a disproportionate number of data values above the inhibition cutoffs of 50, 70 and 90. Of the 40 known compounds with an inhibition greater than 90%, 28 of them are found in Region 1 and 15 of these were not used in the reordering. Thus, if one were to synthesize the 321 unknown or unsampled compounds in Region 1,then *at least* 15 of these compounds would have an inhibition greater than 90%. The stochastic optimization method is also able to reorder the compounds so that the high inhibition values are clustered into a small subset of the matrix [24]. However, the orderings in other regions of the matrix are less similar to those revealed by the deterministic optimization method.

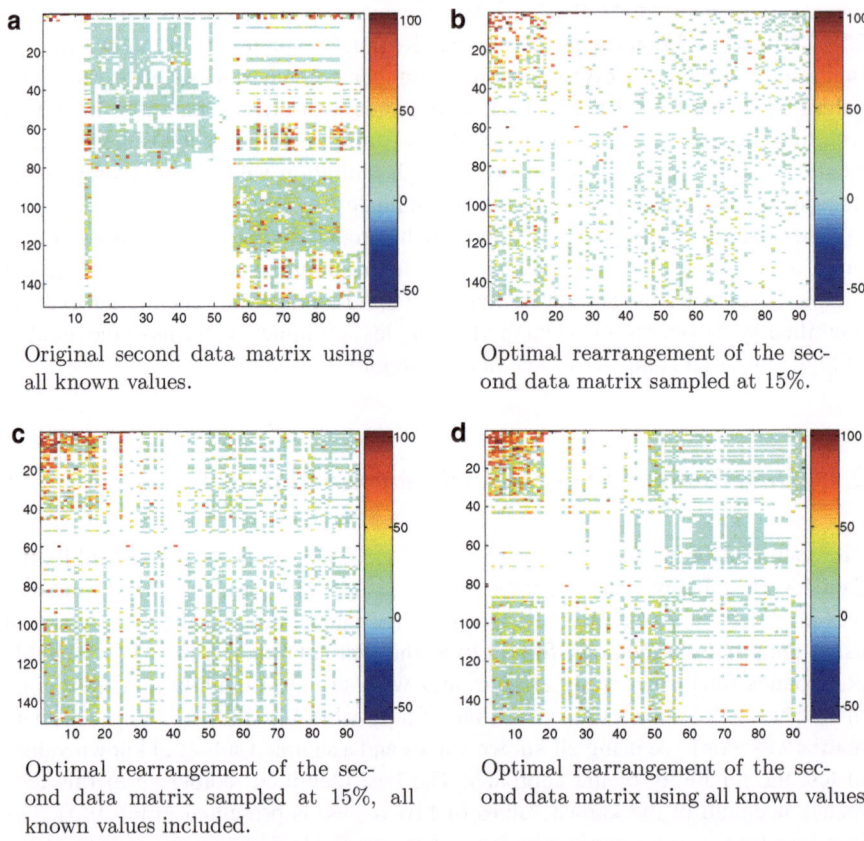

a Original second data matrix using all known values.

b Optimal rearrangement of the second data matrix sampled at 15%.

c Optimal rearrangement of the second data matrix sampled at 15%, all known values included.

d Optimal rearrangement of the second data matrix using all known values.

Fig. 9 Second data matrix reordered by deterministic optimization

2.8 Iterative Synthesis Strategy

Given the optimal ordering of rows and columns, we need to develop a strategy for molecular discovery. For a smooth and regular property landscape, simple local interpolation measures can be applied to represent the expected property of a compound. Such measures assume that the property value of any compound is similar to the average property values of its neighbors. However, a couple of issues come up by using such a strategy. First, when defining average properties, we may want to weight the contributions of neighbors by their distance from a given compound. For any two points (i, j) and (i', j') in the compound property landscape, Euclidean distance is defined by Eq. (43).

$$d_{i',j'}^{i,j} = \sqrt{(i - i')^2 + (j - j')^2}. \tag{43}$$

Furthermore, the absence of information on certain compounds surrounding any given compound has to be accounted for. The set of compounds that will be used to determine the estimated property value of a compound can be represented by $R^{ij}_{i'j'}$, and is given by:

$$R^{ij}_{i'j'} = \left\{ (i', j') : d^{i,j}_{i'j'} \leq d^{thresh} \text{ and } \left(a_{i',j'} \text{ is known or } \left(i = i' \text{ and } j = j' \right) \right) \right\}. \tag{44}$$

The threshold distance, d^{thresh}, is the maximum distance a neighboring point can be from the specified compound (Here, we have used the relation $d^{thresh} = (|I| \cdot 0.1 + |J| \cdot 0.1)/2$. Once we have defined $R^{ij}_{i'j'}$, Eq. (45) is used to define a normalization factor, $\Omega_{i,j}$. This normalization factor provides a higher weight to the closer neighbors and always retains the weight for the compound (i, j) to avoid long-range effects for missing values in sparsely populated regions.

$$\Omega_{i,j} = \sum_{(i',j') \in R^{ij}_{i'j'}} \frac{1}{d^{i,j}_{i'j'} + 1} \tag{45}$$

The estimated property value at (i, j), $\rho_{i,j}$, is then defined by Eq. (46), which is a normalized average of the weighted neighboring property values.

$$\rho_{i,j} = \sum_{(i',j'):a_{i',j'} \text{ is known}} \frac{a_{i',j'}}{\left(d^{i,j}_{i'j'} + 1 \right) \cdot \Omega_{i,j}}. \tag{46}$$

One strategy for synthesis, given an initial sampling of compound property values and its associated optimal substituent ordering, is to sort the estimated property values, $\rho_{i,j}$, and synthesize some number of compounds with the highest predicted property values. As more property values are determined, the process of reordering the substituents and estimating the property values can be repeated in an iterative fashion. An important feature to examine for such an iterative strategy is how much initial data is required to synthesize the important compounds while simultaneously minimizing the number of total compounds synthesized. To test the utility of such a protocol, we have developed an algorithm which begins with a small population of known compounds and uses the reordering results to perform an "in silico" synthesis. After each reordering, Eq. (46) is used to estimate the inhibition values for the unknown compounds. We then select the top 50 unknown compounds of highest predicted inhibition for in silico synthesis (i.e., reveal their actual values) and subsequently reorder this new library. A representative flow diagram for the iterative algorithm is presented in Fig. 10.

This procedure was applied to four sparse data matrices: samplings of 50% (2,050 compounds or 15% of the whole library space), 25% (1,025 compounds

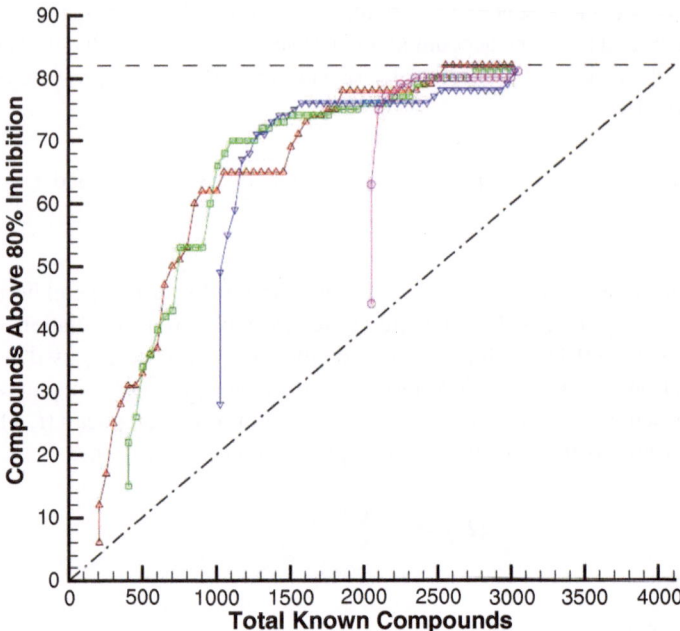

Fig. 10 Iterative strategy for different initial populations of second data matrix for 80% inhibition. The *purple, blue, green* and *red curves* represent starting with 50, 25, 10, and 5% of the available data, respectively

or 7.3% of the whole library space), 10% (411 compounds or 3% of the whole library space) at random from the second data matrix (4,110 known values), and 5% (206 compounds or 1.5% of the whole library space) at random from the original data matrix (4,110 known values). The horizontal dashed line at the top of each figure denotes the total number of compounds that are above that percent inhibition in the original data. The diagonal dashed-dotted line in each figure represents the average gain per synthesis for the original synthesis procedure based on the data provided by Pfizer. To assess the overall effectiveness of the iterative strategy for finding higher inhibition compounds, we counted the total number of compounds above 40–90% inhibition in increments of 10%.

The results for compounds above 80% inhibition are shown in Fig. 11. In this figure, we see that the gain from each of the synthesis curves is fairly consistent among the initial samplings. Each of the curves exhibits a sharp slope in the beginning, indicating a good yield of higher inhibition compounds per synthesis iteration of 50 compounds. It is important to highlight that the synthesis curve starting from only 5% of the original data (206 known values) achieves a better yield of high inhibition compounds than all the other curves when 3,000 known compounds are revealed (see the red curve in Fig. 11).

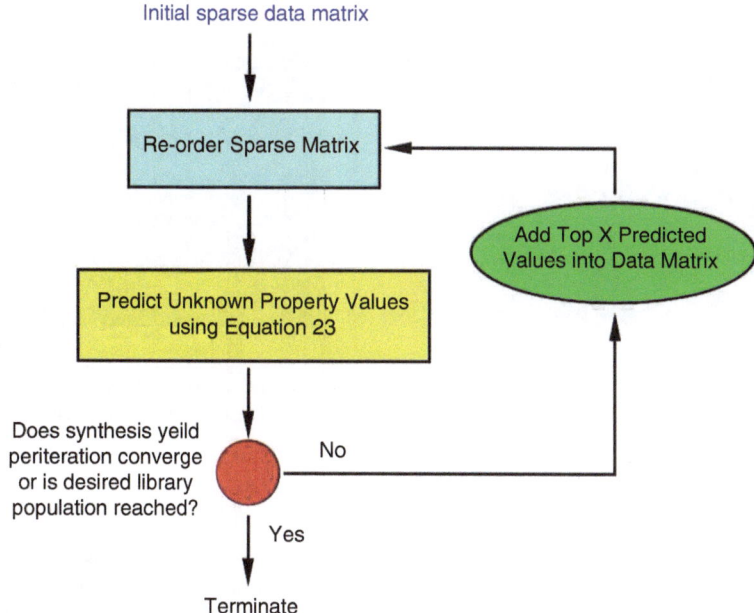

Fig. 11   Flow diagram for iterative algorithm

3   Prediction of In Vivo Chemical Toxicity

A recent application of the dense and sparse matrix clustering algorithms, in conjunction with logistic regression, was to predict the in vivo toxicity of chemicals using their measured in vitro assay data. Development of methods which can screen thousands of industrial and agricultural chemicals has been a major recent initiative in predictive toxicology [42]. The available toxicology data for modeling is typically biased towards toxic chemicals. Further, lack of resources preclude the extensive testing of chemicals. For this analysis, all of the data is available from the EPA ToxCast web site http://www.epa.gov/ncct/toxcast. The in vitro dataset consists of 615 assays (including a set of biochemical receptor and enzyme assays, as well as eight cell-based assays measuring RNA and protein, cytotoxicity, cell growth, and morphology changes) in the form of AC50 and LEC values for a library of 309 chemicals. The in vivo toxicity data provided toxicity data from chronic/cancer rat and cancer mouse studies, multi-generational reproduction rat studies and prenatal developmental toxicity studies in rats and rabbits for 309 chemicals. However, only 78.3% of all possible values were available, thus creating a sparse matrix. Further details on the specific type of data provided in the in vitro and in vivo datasets are available elsewhere [23].

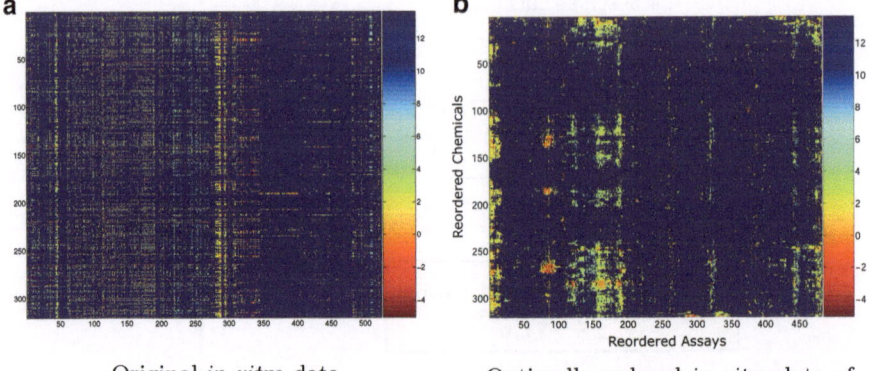

Original *in vitro* data.   Optimally ordered *in vitro* data af-
                                             ter row and column clustering.

Fig. 12 Original and re-ordered in vitro data matrix

After removing the assays which did not show any deviation across the 320 chemicals in the in vitro dataset, we carried out dense matrix clustering on the matrix of 524 rows and 320 columns. The original and re-ordered matrix (after both row and column clustering) is presented in Fig. 12(a), b.

Looking more specifically at the results, it is seen that cytochrome P450 assays come together into two main clusters. In addition, a significant grouping of nuclear receptors was also observed. Along the re-ordered assay dimension, it was observed that specific assay technologies tended to group together. Additional details and figures associated with the dense matrix clustering results can be found in recent work by DiMaggio et al. [23].

The in vivo data matrix provided contains 76 continuous endpoints and 348 binary endpoints. The original data matrix containing the 76 continuous endpoints is shown in Fig. 13a. The optimally re-ordered matrix using the MILP formulation described in the previous section is shown in Fig. 13b.

On further analysis of the clustered endpoints, we observe a physiological-based clustering of endpoints that can be grouped as "reproductive" (containing the words "maternal", "pregnancy", "lactation", "litter", "fetal", "fertility", "ovary", "mating", or "uterus") and "liver". Further, the endpoints in the "reproductive" category were typically from developmental rabbit or multigenerative rat experiments, while the "liver" endpoints were primarily from chronic rat and mouse sources. In a similar manner, the data matrix containing 348 binary endpoints was clustered, and the original and final matrices are shown in Fig. 14a, b respectively. On observation of the endpoints clustering for this data matrix, we see chronic mouse endpoints between positions 1 and 179, while chronic rat endpoints are in positions 180 through 348. There is also a significant physiological clustering of the chronic binary endpoints containing the "liver" descriptor, where three of the chronic mouse endpoints placed in the chronic rat-rich region are associated with these endpoints.

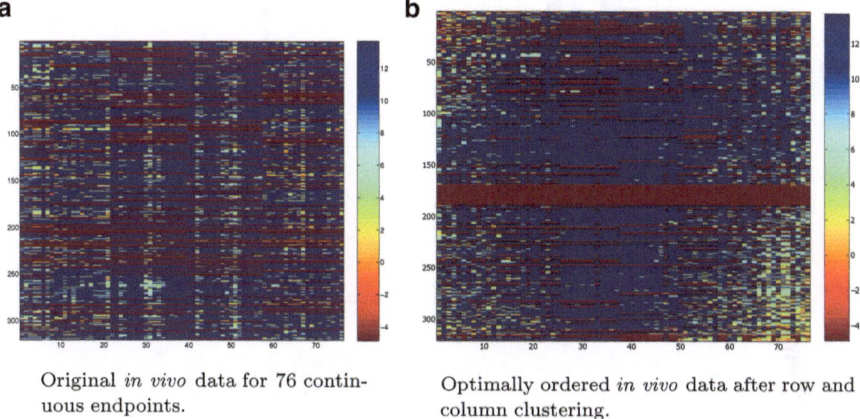

Original *in vivo* data for 76 contin-   Optimally ordered *in vivo* data after row and
uous endpoints.   column clustering.

Fig. 13 Original and re-ordered continuous in vivo data matrix

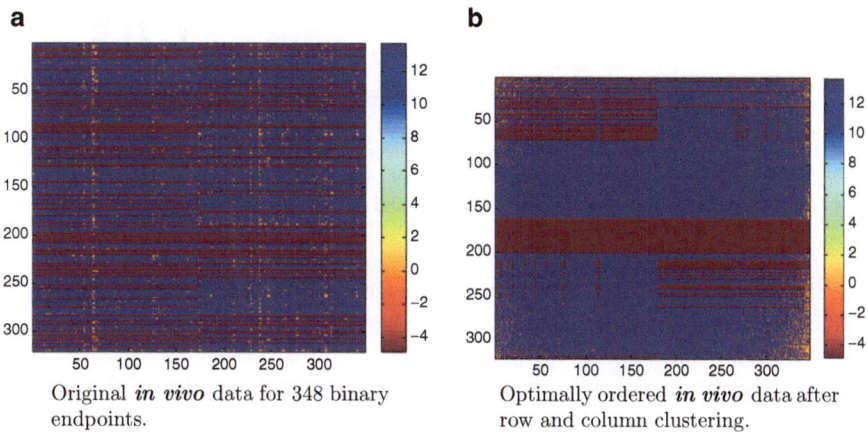

Original *in vivo* data for 348 binary   Optimally ordered *in vivo* data after
endpoints.   row and column clustering.

Fig. 14 Original and re-ordered binary in vivo data matrix

3.1 *Feature Selection Using Logistic Regression*

In an attempt to identify the minimum set of in vitro assay data required to perfectly
classify chemicals as toxic or non-toxic, we started with the set of endpoints
identified as belonging to the "reproductive" and "liver" categories. For each
endpoint, starting with an initial set of 400 descriptors, a rank-and-drop strategy was
used to eliminate descriptors which are not contributing any additional information
to the model. By maximizing the log likelihood of the data given, and with the
introduction of a quadratic regularization term, we identify relative weights for
each of the descriptors in the active set for any iteration. After each iteration, the
standard error associated with each feature is computed by inverting the Hessian

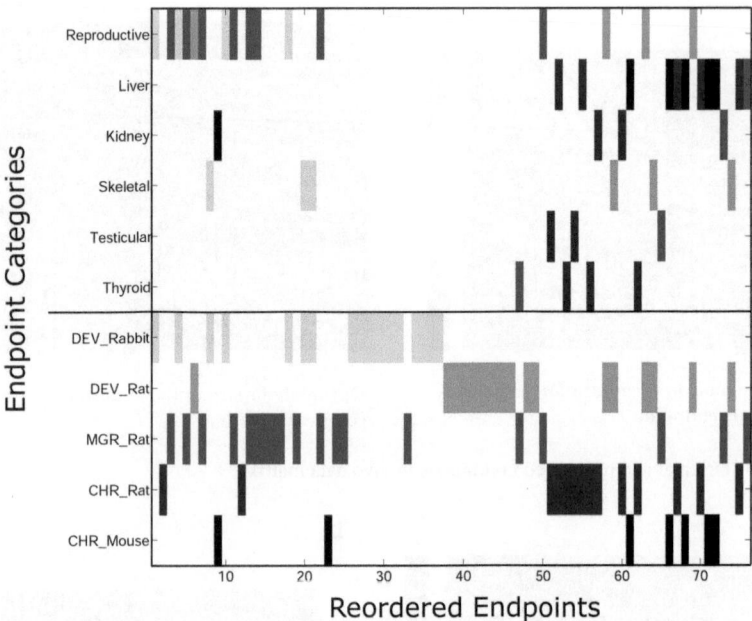

Fig. 15 Relative clustering of related endpoints. The *shaded* elements indicate the existence of a particular endpoint in the given position. Significant clustering is observed with respect to physiological category (i.e., liver and reproduction) and animal species

matrix of the log likelihood function, and the ten features with the lowest parameter value to standard error ratio are eliminated. This procedure is continued until perfect predictions no longer take place [23].

Since the liver and reproductive clusters of endpoints were observed to exhibit anti-correlative behavior (see Fig. 15), one should expect that the selected in vitro descriptors are consistent within the liver or reproductive clusters, but also significantly different between liver and reproductive clusters.

To highlight the differences in the type of descriptors selected between the liver and reproductive in vivo clusters, we computed two fractions corresponding to the relative number of times a particular in vitro assay was determined to be significant in the set of liver and reproductive endpoints, respectively. These two fractions were then sorted based on their absolute differences (i.e., the absolute difference between the relative number of times an in vitro descriptor is selected as significant for a liver associated endpoint and the relative number of times it is selected as significant for a reproductive endpoint), since a larger absolute difference implies the an in vitro has a specificity for either the liver or reproductive endpoints. The most significant differences are presented in Fig. 16.

It is interesting to note in Fig. 16 that the liver endpoints are shown to preferably select cytochrome P450 assays corresponding to subfamily "A" (i.e., 1 CYP1A and 2 CYP3A) when compared to the reproductive endpoints. This makes biological

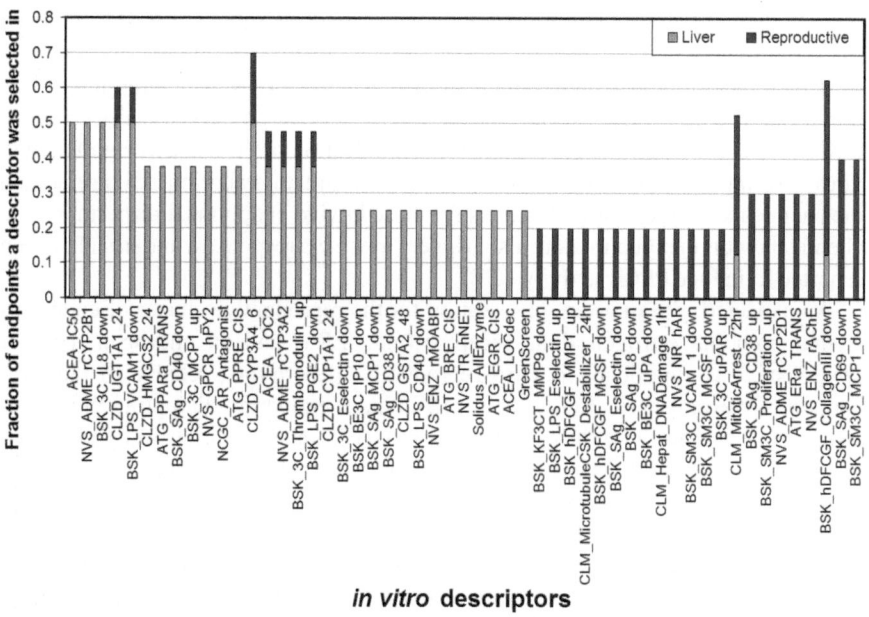

Fig. 16 Difference between descriptors selected for the liver versus reproductive related endpoints

sense, as CYP1A is induced by a number of xenobiotics [20] and CYP3A enzymes are very active in steroid and bile acid 6β-hydroxylation and the oxidation of many xenobiotics [36]. Interestingly, CYP3A has a wide substrate specificity, is prominently expressed in the liver, and is among the most important group of enzymes involved in drug metabolism [80]. CYP2B, which is also preferably selected by the liver endpoints as seen on the left side of Fig. 16, is a large gene family and the regulation of some isoforms is strongly induced by a structurally diverse array of xenobiotics, including pesticides [36]. Many of the nuclear receptors are known to be affected by the CYP2B substrate/product. As shown in Fig. 16, three out of the seven total real time cell electronic sensing assays (ACEA), which measure general cytotoxicity in terms of changes in cell growth kinetics, are also determined to be significant descriptors for the liver endpoints. It should be noted that these assays are found in different biclusters, so their responses over the chemicals are fairly distinct. Lastly, we observe in Fig. 16 that certain nuclear receptors that are well-known regulators of cytochrome P450 genes are selected as specific to the liver cluster of endpoints. These in vitro descriptors include PPARa, AR-agonist, and PPRE. It should be noted here that a recent study screened a set of 200 pesticides for peroxisome proliferator-activated receptor (PPAR) activity (which is expressed in the liver, heart, muscle and kidney) by specifically targeting the receptor activities of PPARa and PPARg [75]. The agonistic activities of the pesticides were measured in relative effective concentrations (REC) to some standard. In this study, it was found that the chemicals diclofop-methyl and imazalil (which notably have very different chemical structures) showed PPARa mediated transcriptional activities, and the in vivo effects of diclofop-methyl and imazalil were then measured by examining the induction of CYP4A gene expression. It was found that diclofop-methyl also induced high levels of CYP4A10 and CYP4A14 mRNA. These findings are consistent with the selection of PPARa as a significant in vitro descriptor for the liver cluster, since diclofop-methyl and imazalil are among the chemicals with lowest LEL values for these in vivo endpoints. Furthermore, the chemical diethylhexyl phthalate also triggers low LEL responses, and has also been reported to induce PPAR activity [37].

An expected, yet assuring, observation for the reproductive cluster of endpoints is that both the estrogen-alpha receptor (e.g., ERa) and an androgen receptor are selected as significant in vitro reproductive descriptors, relative to the liver endpoints. Several agricultural chemicals contain endocrine-disrupting properties through interactions with the estrogen receptor (ER), and an earlier study identified 80 out of 200 chemicals as having ER receptor activity [51]. Our results are found to be consistent with these previous findings, which reported 34 pesticides displaying both ER and antiAR activity [51].

Within both the liver and reproductive endpoints, it was observed that several clusters of differentiations, including CD38, CD40, CD69, and CD141 (thrombomodulin), were selected as significant descriptors. These clusters of differentiations are known to be important factors in immune response. Consistent with these assays are the selection of descriptors associated with chemokines, which attract leukocytes to infection sites. They are assigned into four different groups based upon their conserved cysteine residues: C-C, C-X-C, C, and CX3C. In Fig. 16, it is seen that three CC motif (MCP-1) and a C-X-C motif (IP-10, which is secreted in response to INF-γ) chemokines are selected by the reproductive and liver endpoints.

4   Conclusions

In this chapter, we have presented rigorous methods for the optimal re-ordering of data matrices. For the re-ordering of dense matrices, we have presented our clustering method OREO, which can be established either as a network flow model, or as a representation of the traveling salesman problem. An iterative approach can be used to divide the original data matrix into subclusters, and carry out similar objectives on each subcluster to get further classification of rows and columns. The performance of OREO was evaluated on a number of biological systems like metabolite concentration data, colon cancer data, breast cancer data and yeast segregation data. The applicability of the method to non-biological fields was demonstrated through the performance of OREO on image reconstruction data. Analysis of the performance of OREO on these datasets showed an improved agglomeration of related metabolites and annotated genes, thus representing the advantage of the global optimal re-ordering algorithm over local solutions.

For the sparse clustering problems, we have introduced a mixed-integer linear programming (MILP) based formulation which aims to bring rows and columns together, by accounting for their positions in the entire matrix. The performance of the method was evaluated on a moderate-sized and a large-sized data matrix provided by Pfizer, Inc. The resulting clusters showed a good collection of data values with a high degree of inhibition in a small subset of the original matrix, thus showing the efficacy of the re-ordering algorithm.

Finally, the dense and sparse clustering algorithms were applied to toxicology data, with the aim of developing a method for predicting the in vivo toxicity of chemicals, given in vitro assay data. Results from the application of the dense clustering algorithm showed that similar assays were grouped together, as were assays related to similar technologies. The sparse clustering algorithm showed a clustering of "reproductive" and "liver" endpoints. Logistic regression was used to identify a small subset of in vitro assays for the toxicity prediction of chemicals for all reproductive and liver endpoints. A good correlation between cytochrome P450 family 'A' and the liver endpoints was observed. Similarly, estrogen and androgen receptors were favorably preferred selections for the reproductive endpoints.

Acknowledgements CAF gratefully acknowledges financial support from the National Science Foundation, National Institutes of Health (R01 GM52032; R24 GM069736) and U.S. Environmental Protection Agency EPA (GAD R 832721-010).

References

1. A. Aggarwal, C.A. Floudas, Synthesis of general separation sequences - nonsharp separations. Comp. Chem. Eng. **14**(6), 631–653 (1990)
2. U. Alon, N. Barkai, D.A. Notterman, K. Gish, S. Ybarra, D. Mack, A.J. Levine, Broad patterns of gene expression revealed by clustering analysis of tumor and normal colon tissues probed by oligonucleotide arrays. Proc. Natl. Acad. Sci. **96**, 6745–6750 (1999)
3. M.R. Anderberg, *Cluster Analysis for Applications* (Academic, New York, 1973)
4. I.P. Androulakis, C.D. Maranas, C.A. Floudas, Prediction of oligopeptide conformations via deterministic global optimization. J. Glo. Opt. **11**, 1–34 (1997)
5. D.L. Applegate, R.E. Bixby, V. Chvatal, W.J. Cook, *The Traveling Salesman Problem: A Computational Study* (Princeton University Press, Princeton, 2007)
6. P. Armutlu, M.E. Ozdemir, F. Uney-Yuksektepe, I.H. Kavakli, M. Turkay, Classification of drug molecules considering their ic50 values using mixed-integer linear programming based hyper-boxes method. BMC Bioinformatics **9**, 411 (2008)
7. W. Bannwarth, B. Hinzen, R. Mannhold, H. Kubinyi, G. Folkers, *Combinatorial Chemistry: From Theory to Application (Methods and Principles in Medicinal Chemistry)* (Wiley, New Jersey, 2006)
8. Z. Bar-Joseph, E.D. Demaine, D.K. Gifford, N. Srebro, A.M. Hamel, T.S. Jaakola, K-ary clustering with optimal leaf ordering for gene expression data. Bioinformatics **19**(9), 1070–1078 (2003)
9. J.N. Bhuyan, V.V. Raghavan, K.E. Venkatesh, in *Genetic Algorithm for Clustering with an Ordered Representation.* Proceedings of the Fourth International Conference on Genetic Algorithms, p. 408–415 (1991)

10. S. Bleuler, A. Prelic, E. Zitzler, *An EA Framework for Biclustering of Gene Expression Data.* IEEE Congress on Evolutionary Computation, pp. 166–173 (2004)

11. M. J. Brauer, J. Yuan, B. Bennett, W. Lu, E. Kimball, D. Bostein, J.D. Rabinowitz, Conservation of the metabolomic response to starvation across two divergent microbes. Proc. Natl. Acad. Sci. **103**, 19302–19307 (2006)

12. R.B. Brem, L. Kruglyak, The landscape of genetic complexity across 5,700 gene expression traits in yeast. Proc. Natl. Acad. Sci. **102**(5), 1572–1577 (2005)

13. S. Busygin, O.A. Prokopyev, P.M. Pardalos, Feature selection for consistent biclustering via fractional 0-1 programming. J. Comb. Opt. **10**, 7–21 (2005)

14. S. Busygin, O.A. Prokopyev, P.M. Pardalos, An optimization based approach for data classification. Opt. Meth. Soft. **22**(1), 3–9 (2007)

15. P. Carmona-Saez, R.D. Pasqual-Marqui, F. Tirado, J. Carazo, A. Pascual-Montano, Biclustering of gene expression data by non-smooth non-negative matrix factorization. BMC Bioinformatics **7**, 78–96 (2006)

16. Y. Cheng, G.M. Church, Biclustering of expression data. Proc. ISMB 2000, pp. 93–103 (2000)

17. A.R. Ciric, C.A. Floudas, A retrofit approach for heat-exchanger networks. Comp. Chem. Eng. **13**(6), 703–715 (1989)

18. S. Climer, W. Zhang, Rearrangement clustering: Pitfalls, remedies, and applications. J. Mach. Learn. Res. **7**, 919–943 (2006)

19. CPLEX, ILOG CPLEX 9.0 User's Manual (2005)

20. M.S. Denison, J.P. Whitlock, Xenobiotic-inducible transcription of cytochrome P450 genes. J. Biol. Chem. **270**(31), 18175–18178 (1995)

21. P. DiMaggio, S. McAllister, C.A. Floudas, X.J. Feng, J. Rabinowitz, H. Rabitz, Biclustering via optimal re-ordering of data matrices in systems biology: Rigorous methods and comparative studies. BMC Bioinformatics **9**, 458 (2008)

22. P. DiMaggio, S. McAllister, C.A. Floudas, X.J. Feng, J. Rabinowitz, H. Rabitz, A network flow model for biclustering via optimal re-ordering of data matrices. J. Glo. Opt. **47**, 343–354 (2010)

23. P.A. DiMaggio, A. Subramani, R.S. Judson, C.A. Floudas, A novel framework for predicting *in vivo* toxicities from in vitro data using optimal methods for dense and sparse matrix reordering and logistic regression. Toxicol. Sci. **118**, 251–265 (2010)

24. P.A. DiMaggio, S.R. McAllister, C.A. Floudas, X.J. Feng, J.D. Rabinowitz, H.A. Rabitz, Enhancing molecular discovery using descriptor-free rearrangement clustering techniques for sparse data sets. AIChE J **56**, 405–418 (2010)

25. F. Divina, J. Aguilar, Biclustering of expression data with evolutionary computation. IEEE Trans. Knowl. Data Eng. **18**(5), 590–602 (2006)

26. A.W.F. Edwards, L.L. Cavalli-Sforza, A method for cluster analysis. Biometrics **21**, 362–375 (1965)

27. M.B. Eisen, P.T. Spellman, P.O. Brown, D. Botstein, Cluster analysis and display of genome-wide expression patterns. Proc. Natl. Acad. Sci. **95**, 14863–14868 (1998)

28. C.A. Floudas, *Nonlinear and Mixed-Integer Optimization* (Oxford University Press, New York, 1995)

29. C.A. Floudas, S.H. Anastasiadis, Synthesis of distillation sequences with several multicomponent feed and product streams. Chem. Eng. Sci. **43**(9), 2407–2419 (1988)

30. C.A. Floudas, I.E. Grossmann, Synthesis of flexible heat exchanger networks with uncertain flowrates and temperatures. Comp. Chem. Eng. **11**(4), 319–336 (1987)

31. L.R. Ford, D.R. Fulkerson, *Flows in Networks* (Princeton University Press, NJ, 1962)

32. H.K. Fung, C.A. Floudas, M.S. Taylor, L. Zhang, D. Morikis, Towards full sequence de novo protein design with flexible templates for human beta-defensin-2. Biophys. J. **94**, 584–599 (2008)

33. C. Hansch, A. Leo, *Exploring QSAR – Fundamentals and Applications in Chemistry and Biology* (American Chemical Society, Washington, DC, 1995)

34. C. Hansch, B.R. Telzer, L. Zhang, Comparative qsar in toxicology: Examples from teratology and cancer chemotherapy of aniline mustards. Crit. Rev. Toxicol. **25**, 67–89 (1995)

35. J.A. Hartigan, M.A. Wong, Algorithm AS 136: A K-means clustering algorithm. Appl. Stat. **28**, 100–108 (1979)
36. P. Honkakoski, M. Negishi, Regulation of cytochrome P450 (CYP) genes by nuclear receptors. Biochem. J. **347**, 321–337 (2000)
37. W.W. Huber, B. Grasl-kraupp, R. Schulte-hermann, Hepatocarcinogenic potential of di(2-ethylhexyl)phthalate in rodents and its implications on human risk. Crit. Rev. Toxicol. **26**(4), 365–481 (1996)
38. J. Huser, R. Mannhold, H. Kubinyi, G. Folkers, *High-Throughput Screening in Drug Discovery (Methods and Principles in Medicinal Chemistry)* (Wiley-VCH, NJ, 2006)
39. A.K. Jain, P.J. Flynn, in *Image Segmentation Using Clustering*, ed. by N. Ahuja, K. Bowyer. Advances in Image Understanding: A Festschrift for Azriel Rosenfeld (IEEE, NJ, 1996), pp. 65–83
40. A.K. Jain, J. Mao, Artificial neural networks: A tutorial. IEEE Comp. **29**, 31–44 (1996)
41. S.L. Janak, X. Lin, C.A. Floudas, Enhanced continuous-time unit-specific event based formulation for short-term scheduling of multipurpose batch processes: Resource constraints and mixed storage policies. Ind. Eng. Chem. Res. **43**, 2516–2533 (2004)
42. R. Judson, A. Richard, D.J. Dix, K. Houck, M. Martin, R. Kavlock, V. Dellarco, T. Henry, T. Holderman, P. Sayre, S. Tan, T. Carpenter, E. Smith, The toxicity data landscape for environmental chemicals. Environ. Health Perspect. **117**, 685–695 (2009)
43. P. Kahraman, M. Turkay, Classification of 1,4-dihydropyridine calcium channel antagonists using the hyperbox approach. Ind. Eng. Chem. Res. **46**, 4921–4929 (2007)
44. R.W. Klein, R.C. Dubes, Experiments in projection and clustering by simulated annealing. Pattern Recogn. **22**, 213–220 (1989)
45. J.L. Klepeis, C.A. Floudas, Free energy calculations for peptides via deterministic global optimization. J. Chem. Phys. **110**, 7491–7512 (1999)
46. J.L. Klepeis, C.A. Floudas, Ab initio tertiary structure prediction of proteins. J. Glo. Opt. **25**, 113–140 (2003)
47. J.L. Klepeis, C.A. Floudas, ASTRO-FOLD: A combinatorial and global optimization framework for ab initio prediction of three-dimensional structures of proteins from the amino acid sequence. Biophys. J. **85**, 2119–2146 (2003)
48. J.L. Klepeis, C.A. Floudas, D. Morikis, J.D. Lambris, Predicting peptide structures using NMR data and deterministic global optimization. J. Comp. Chem. **20**(13), 1354–1370 (1999)
49. J.L. Klepeis, C.A. Floudas, D. Morikis, C.G. Tsokos, E. Argyropoulos, L. Spruce, J.D. Lambris, Integrated computational and experimenal approach for lead optimization and design of compstatin variants with improved activity. J. Am. Chem. Soc. **125**(28), 8422–8423 (2003)
50. Y. Kluger, R. Basri, J.T. Chang, M. Gerstein, Spectral biclustering of microarray data: Coclustering genes and conditions. Genome Res. **13**, 703–716 (2003)
51. H. Kojima, E. Katsura, S. Takeuchi, K. Niiyama, K. Kobayashi, Screening for estrogen and androgen receptor activities in 200 pesticides by in vitro reporter gene assays using chinese hamster ovary cells. Environ. Health Perspect. **112**(5), 524–531 (2004)
52. A.C. Kokossis, C.A. Floudas, Optimization of complex reactor networks-II: nonisothermal operation. Chem. Eng. Sci. **49**(7), 1037–1051 (1994)
53. J.K. Lenstra, Clustering a data array and the traveling-salesman problem. Oper. Res. **22**(2), 413–414 (1974)
54. J.K Lenstra, A.H.G. Rinnooy Kan, Some simple applications of the traveling-salesman problem. Oper. Res. Q. **26**(4), 717–733 (1975)
55. F. Liang, X. Feng, M. Lowry, H. Rabitz, Maximal use of minimal libraries through the adaptive substituent reordering algorithm. J. Phys. Chem. B **109**, 5842–5854 (2005)
56. X. Lin, C.A. Floudas, Design, synthesis and scheduling of multipurpose batch plants via an effective continuous-time formulation. Comp. Chem. Eng. **25**, 665–674 (2001)
57. M. Lutz, T. Kenakin, *Quantitative Molecular Pharmacology and Informatics in Drug Discovery* (Wiley, NJ, 2001)
58. S.C. Madeira, A.L. Oliveira, Biclustering algorithms for biological data analysis: A survey. IEE-ACM Trans. Comp. Bio. **1**(1), 24–45 (2004)

59. W.T. McCormick Jr., P.J. Schweitzer, T.W. White, Problem decomposition and data reorganization by a clustering technique. Oper. Res. **20**(5), 993–1009 (1972)

60. M. Mönnigmann, C.A. Floudas, Protein loop structure prediction with flexible stem geometries. Protein Struct. Funct. Bioinformatics **61**, 748–762 (2005)

61. P. Moscato, A. Mendes, R. Berretta, Benchmarking a Memetic algorithm for ordering microarray data. Biosystems **88**(1), 56–75 (2007)

62. R. Ng, *Drugs – From Discovery to Approval* (WileyLiss, NJ, 2006)

63. P.M. Pardalos, V. Boginski, A. Vazakopoulos, *Data Mining in Biomedicine* (Springer, Berlin, 2007)

64. R. Perkins, H. Fang, W. Tong, W. Welsh, Quantitative structure-activity relationship methods: perspectives on drug discovery and toxicology. Environ. Toxicol. Chem. **22**, 1666–1679 (2003)

65. A. Prelic, S. Bleuler, P. Zimmermann, A. Wille, P. Buhlmann, W. Gruissem, L. Hennig, L. Thiele, E. Zitzler, A systematic comparison and evaluation of biclustering methods for gene expression data. Bioinformatics **22**(9), 1122–1129 (2006)

66. V.V. Raghavan, K. Birchand, in *A Clustering Strategy Based on a Formalism of the Reproductive Process in a Natural System*. Proceedings of the Second International Conference on Information Storage and Retrieval, pp. 10–22 (1979)

67. D.J. Reiss, N.S. Baliga, R. Bonneau, Integrated biclustering of heterogeneous genome-wide datasets for the inference of global regulatory networks. BMC Bioinformatics **7**, 280–302 (2006)

68. G. Salton, Developments in automatic text retrieval. Science **253**, 974–980 (1991)

69. N. Shenvi, J.M. Geremia, H. Rabitz, Substituent ordering and interpolation in molecular library optimization. J. Phys. Chem. **107**, 2066–2074 (2003)

70. N. Shenvi, J.M. Geremia, H. Rabitz, Substituent ordering and interpolation in molecular library optimization. J. Phys. Chem. A **107**, 2066 (2003)

71. H.D. Sherali, J. Desai, A global optimization RLT-based approach for solving the fuzzy clustering problem. J. Glo. Opt. **33**, 597–615 (2005)

72. H.D. Sherali, J. Desai, A global optimization RLT-based approach for solving the hard clustering problem. J. Glo. Opt. **32**, 281–306 (2005)

73. N. Slonim, G.S. Atwal, G. Tkacik, W. Bialek, Information-based clustering. Proc. Natl. Acad. Sci. **102**(51), 18297–18302 (2005)

74. A. Subramani, P.A. DiMaggio Jr., C.A. Floudas, Selecting high quality structures from diverse conformational ensembles. Biophys. J. **97**, 1728–1736 (2009)

75. S. Takeuchi, T. Matsuda, S. Kobayashi, T. Takahashi, H. Kojima, In vitro screening of 200 pesticides for agonistic activity in mouse peroxisome proliferator-activated receptor PPARa and PPARg and quantitative analysis of in vivo induction pathway. Toxicol. Appl. Pharmacol. **217**, 235–244 (2008)

76. M.P. Tan, J.R. Broach, C.A. Floudas, A novel clustering approach and prediction of optimal number of clusters: Global optimum search with enhanced positioning. J. Glo. Opt. **39**, 323–346 (2007)

77. M.P. Tan, J.R. Broach, C.A. Floudas, Evaluation of normalization and pre-clustering issues in a novel clustering approach: Global optimum search with enhanced positioning. J. Bioin. Comp. Bio **5**(4), 895–913 (2007)

78. M.P. Tan, E. Smith, J.R. Broach, C.A. Floudas, Microarray data mining: A novel optimization-based approach to uncover biologically coherent structures. BMC Bioinformatics **9**, 268–283 (2008)

79. A. Tanay, R. Sharan, R. Shamir, Discovering statistically significant biclusters in gene expression data. Bioinformatics **18**, S136–S144 (2002)

80. L.E. Thummel, G.R. Wilkinson, In vitro and in vivo drug interactions involving human CYP3A. Annu. Rev. Pharmacol. Toxicol. **38**, 389–430 (1998)

81. W. Tong, W. Welsh, L. Shi, H. Fang, R. Perkins, Structure-activity relationship approaches and applications. Environ. Toxicol. Chem. **22**, 1680–1695 (2003)

82. H.L. Turner, T.C. Bailey, W.J. Krzanowski, C.A. Hemingway, Biclustering models for structured microarray data. IEEE/ACM Trans. Comput. Biol. Bioinformatics **2**(4), 316–329 (2005)

83. L.J. van't Veer, H. Dai, M.J. Vijver, Y.D. He, A.A. Hart, M. Mao, H.L. Peterse, K. van der Kooy, M.J. Marton, A.T. Witteveen, G.J. Schreiber, R.M. Kerkhoven, C. Roberts, P.S. Linsley, R. Bernards, S.H. Friend, Gene expression profiling predicts clinical outcome of breast cancer. Nature **415**, 530–536 (2002)
84. J.H. Wolfe, Pattern clustering by multivariate mixture analysis. Multivariate Behav. Res. **5**, 329–350 (1970)
85. S. Yoon, C. Nardini, L. Benini, G. De Micheli, Discovering coherent biclusters from gene expression data using zero-suppressed binary decision diagrams. IEEE/ACM Trans. Comput. Biol. Bioinformatics **2**(4), 339–354 (2005)
86. Y. Zhang, J. Skolnick, SPICKER: A clustering approach to identify near-native protein folds. J. Comput. Chem. **25**, 865–871 (2004)

Clustering Time Series Data with Distance Matrices

Onur Şeref and W. Art Chaovalitwongse

Abstract Clustering is a frequently used method in unsupervised analysis of various data types including time series data. In this study, we first present a discrete k-median (DKM) method based on an uncoupled bilinear programming algorithm and modify it for faster implementation, which becomes a variant of the Lloyd's algorithm. We also introduce a fuzzy discrete k-median (FDKM) method which is the fuzzy version of the modified algorithm. The main draw for the these two efficient algorithms is that they do not require any input but a matrix of distances as a measure of dissimilarity between pairs of samples to avoid the complications that may arise from working with the actual domain that the data samples reside in. We also include a hiearchical cluster tree (HCT) method and partition around medoids (PAM) method, both of which can use the distance matrix for clustering. The output of all four methods are median samples, which define clusters by assigning each sample to the closest median sample using the distance matrix. We consider four different distance measures, rectilinear, Euclidean, squared-Euclidean and dynamic time warping (DTW) to create the distance matrix, and also mention how the calculation of the distance matrix can be extended to any kernel induced feature space. The main application domain in this study is time series data, where actual samples in the data set are better cluster representations than mean or median points whose components are independently calculated for each dimension of the domain. We present computational results on a public time series

O. Şeref (✉)
Business Information Technology, Virginia Polytechnic Institute
and State University, Blacksburg, VA 24061, USA
e-mail: seref@vt.edu

W.A. Chaovalitwongse
Department of Industrial and Systems Engineering, Department of Radiology,
University of Washington, Seattle, WA 98195, USA
e-mail: artchao@uw.edu

P.M. Pardalos et al. (eds.), *Optimization and Data Analysis in Biomedical Informatics*,
Fields Institute Communications 63, DOI 10.1007/978-1-4614-4133-5_2,
© Springer Science+Business Media New York 2012

benchmark data set and a real life local field potential (LFP) recordings collected from a macaque monkey brain during a visuomotor task.

Mathematics Subject Classification (2010): Primary 62H30, Secondary 68W25

1   Introduction

Clustering is an unsupervised method for data analysis that is concerned with partitioning data into subsets called clusters. In general, a cluster is comprised of samples that are more "similar" among each other than the samples in other clusters. Clustering has been extensively studied both in theory and application [26]. Clustering algorithms can be classified in a number of ways depending on how they represent clusters and how samples are assigned to these clusters. One of the major splits among clustering methods is between *hierarchical* and *partitional* approaches. Hierarchical methods produce a single nested partition, whereas partitional methods lack this structure. Most commonly used hierarchical approach is an agglomerative method, which starts with each data representing a cluster by itself and progresses by iteratively merging clusters to form a tree called dendrogram. Individual clusters are formed by disconnecting or cutting links on this tree and identifying the leaves of the subtrees formed. Partitional algorithms define partitions from the beginning and improve clusters and the representative set of points for these clusters iteratively. In terms of cluster membership, clustering algorithms can be classified as *hard* versus *fuzzy* clustering algorithms. In hard clustering, a data point belongs to a single cluster only. Cluster membership in fuzzy clustering is distributed over all clusters with varying degrees, which can easily be converted to hard clustering by assigning the data to the cluster for which it has the highest membership degree.

1.1   k-Means Versus k-Median

One of the most well-known partitional clustering method is the k-means clustering. The objective of k-means clustering is to find k cluster centers that minimize the sum of the squared-Euclidean distances of the samples from their closest centers. In k-means, the center of a cluster is the centroid of the samples in that cluster, which can be found by averaging the components of all the samples in the cluster for each dimension of the space that the samples reside in. These centers are used to define new clusters by assigning the samples to the center closest to them. The main algorithm alternates between finding new centers and new clusters over a number of iterations until the algorithm converges to a local optimum. This alternating scheme is first developed by Lloyd [23], which is proven to be optimal for centroids as the centers of clusters when squared-Euclidean distance is used [12]. The k-median clustering method is a variation on k-means clustering, in which the medians for each dimension is calculated to find a median point that represents the cluster. In other words, k-means method minimizes the dissimilarity over all clusters with respect to the squared-Euclidean distance, where as k-median minimizes the dissimilarity with respect to the Euclidean distance itself [16]. With this simple modification, it is easy to convert any k-means approach into a k-median approach.

For either method, initial set of points has a significant effect on the final solution since the original clustering problem has a large number of local minima. In [21], it was shown that a good cluster center initialization can improve these algorithms. In [34], a global k-means method is used for choosing the initial clustering centers. Another efficient method in [4] refines initial points by estimating the modes of a distribution, which allows the iterative algorithm to converge to a "better" local optimal solution. In this study, we employ the initialization method used for partition around medoids (PAM) algorithm to find good starting medians [19].

1.2 Discrete Versus Continuous

In many practical cases, the calculations in the original domain of the data may be difficult or impractical. In some other cases, averaging samples or finding medians individually for each dimension may create centers or medians that cannot represent the samples in a cluster. Especially in time series, the data points are highly interdependent since each dimension is a time point. Therefore, processing each dimension separately does not reflect the shape and structure of time series data domain. We only consider discrete methods as the main application focus of this study, where the discrete points are the samples in the data set. Under these assumptions, one of the actual samples in a cluster is considered to be a better representation of the cluster. We refer to clustering methods that are similar to k-means and k-median clustering, but regard actual sample points as cluster medians as discrete k-means and discrete k-median methods, respectively. The discrete k-median problem is also known as the p-center, p-hub or p-median problem in the facility location problem literature [7, 33]. The p-median problem constitutes a larger class of facility location problems known as minisum location allocation problems.

Discrete k-median problems are generally more difficult problems compared to their non-discrete counterparts. Searching for optimal medians among discrete median points instead of centers or medians in a continuous space is a challenging combinatorial problem. However, the advantage behind this problem is that the dissimilarities between discrete samples can be defined in a number of ways without having to calculate any point that does not belong to the input set of samples, and such dissimilarities is sufficient for clustering methods to work. The results of this study can be used, for example, in studying brain data through electrical recordings or network intrusion detection in network data streams, where actual samples are needed to represent the behavior of time series data.

Fast implementations are crucial for the discrete methods to work efficiently, especially with online time series data. Therefore, the exact methods employed in p-median facility location problems may not be necessary [5, 6]. Good solutions are usually more than sufficient from a practical point of view and generally leave a small optimality gap with near optimal cluster centers and cluster assignments. The discrete k-median algorithm presented in this study produces good solutions very efficiently compared to exact methods. The main approach is based on the work in [24], which is a non-discrete algorithm that is limited to 1-norm distances. We modify the uncoupled bilinear program approach [1] used in [24] in two different ways, by considering a discrete version of the problem, and by allowing arbitrary distance measures to be used between samples, including dynamic time warping, which is the most commonly used distance for time series data [2, 20, 36]. We further modify this algorithm with an additional constraint, which has an explicit solution that is identical to the original solution in most cases but makes the algorithm run much faster. After this constraint, our algorithm becomes a discrete version of Lloyd's algorithm. We refer to this fast algorithm as discrete k-median (DKM) for the remainder of the paper.

1.3   Fuzzy k-Median and Other Distance Based Methods

Next, we introduce a fuzzy version of the fast discrete algorithm, which we refer to as fuzzy discrete k-means (FDKM). This algorithm is also known as fuzzy medoids algorithm [17, 27], which has weighted versions [25], and has applications to time series data [11]. There are other algorithms on fuzzy k-median clustering [8, 22]. We include a hierarchical cluster tree method (HCT) which builds a hierarchical tree called *dendrogram* by merging subclusters together based on an inter-cluster distance definition [32]. Once the dendrogram structure is constructed, it can be cut in various ways to form clusters. HCT method finds applications in time series clustering [31]. We also include partition around medoids (PAM), a partition based algorithm that performs exhaustive swaps between median and non-median samples to improve the objective iteratively based on a distance matrix [19, 28], and has applications to time series [14]. We note that, aside from DKM, we use the initialization method of PAM method to find an initial set of medians for FDKM.

1.4   Applications

We apply DKM, FDKM, HCT and PAM methods on the University of California at Riverside (UCR) time series data sets. We consider four different distance measures to form the distance matrix, namely, rectilinear, Euclidean, squared-Euclidean and dynamic time warping (DTW) distances between pairs of samples. Among these distance measures, the squared-Euclidean distance is also a well-known *Bregman divergence* and is equivalent to the discrete k-means method [3, 9]. Therefore, the results with respect to Euclidean versus squared-Euclidean distance can be considered as the comparison of discrete k-median and discrete k-means methods in a geometric space. We compare clustering performance on the training and test data sets using normalized mutual information (NMI) scores. We also present running time results for each method regarding each data set by averaging running times over each distance measure.

1.5 Notation and Organization

For consistency throughout the text, we define the following: All vectors are column vectors. If A is a matrix, A_i is a vector which denotes the ith column of A, and $\overline{A}_j$ is a row vector which denotes the jth row of A. A_{ij} is the jth entry of the column vector A_i, and $\overline{A}_{ji}$ is the ith entry of row vector $\overline{A}_j$, which implies $A_{ij} = \overline{A}_{ji}$. We consider e to be a column vector of ones of appropriate size and *prime* (′) to be the transpose operator. G_j denotes the set of indices of samples that are assigned to cluster j for $j = 1, \ldots, k$, and we assume that these sets are mutually exclusive, that is $G_j \subset \{1, 2, \ldots, n\}$ and $G_j \cap G_l = \emptyset$ for any $j \neq l$. Distance matrix is denoted as D. The set of samples to be clustered is denoted as S, and set of medians is denoted as $M \subset S$. Both S and M are used as the actual samples or their indices interchangeably throughout the text.

The rest of this paper is organized as follows. In Sect. 2, formulations, properties and the algorithms for DKM, FDKM, HCT and PAM methods are introduced. In Sect. 3, the rectilinear, Euclidean, squared-Euclidean, dynamic time warping, and kernel induced distance measures are introduced, performance measures and computational results are presented. Conclusions are drawn in Sect. 4.

2 Clustering Methods that Use Distance Matrices

We present four clustering methods that take a matrix composed of distances between every pair of samples according to an arbitrary distance measure and create clusters based on this matrix. We first introduce an exact bilinear formulation to solve the discrete k-median problem and demonstrate the iterative uncoupled bilinear program approach (UBPA) to find a good local solution. We further show that, with an additional assumption, this approach transforms into Lloyd's algorithm modified for discrete k-median problem. Next, we introduce a fuzzy version of this algorithm by letting cluster membership be defined as a continuous variable. We also review the hierarchical cluster trees with complete linking [32] and the partition around medoids algorithms, both of which can use a distance matrix as input.

2.1 Discrete k-Median Clustering (DKM)

In the discrete k-median clustering problem, we are given a set of samples S and an integer k, and we want to select a set M of k samples from S to be cluster medians to minimize the sum of the distances from the samples in S to their nearest median. This problem is equivalent to the p-median problem in facility location literature, and is known to be $\mathcal{NP}$-hard on general graphs when p is not fixed [15], even for planar graphs with maximum vertex degree of 3 [18]. The problem can be solved in polynomial time when p is fixed, but is computationally expensive for larger values of p.

The first integer programming formulation for this problem is provided in [30], which includes selection variables and allocation variables denoted by the vector $Y_{n\times 1} \in \{0, 1\}$ and the matrix $Z_{n\times n}$, respectively. Selection variables determine cluster centers Y_i and allocation variables determine the cluster assignment of the samples to these centers Z_{ij}. This integer formulation is given in (1), where D is the distance matrix.

$$
\begin{aligned}
\min_{Y,Z} \quad & \sum_{i=1}^{n} D_i \, Z_i \\
\text{s.t.} \quad & e'Z_i = 1 \qquad \text{for} \quad i = 1, \dots, n \\
& Z_i \leq Y \qquad \text{for} \quad i = 1, \dots, n \\
& e'Y = k \\
& Z \in \{0, 1\}^{m \times m} \\
& Y \in \{0, 1\}^{m}.
\end{aligned}
\tag{1}
$$

2.1.1 Uncoupled Bilinear Programming

An alternative exact formulation is developed using bilinear programming following the general framework in [24], and generalizing it to arbitrary distance measures [13]. This bilinear program is solved to a local optimum using an alternating iterative method.

Let $X_{n\times k}$ be a matrix of decision variables such that if sample i is the pth median then $X_{i,p} = 1$ and 0 otherwise, for $i = 1, \dots, n$ and $p = 1, \dots, k$. Then formulation (2) below solves the k-median problem,

$$
\begin{aligned}
\min_{X} \quad & \sum_{i=1}^{n} \min_{p=1,\dots,k} \{D_i' \, X_p\} \\
\text{s.t.} \quad & e' X_p = 1 \qquad \text{for } p = 1, \dots, k \\
& X \in \{0, 1\}^{n \times k},
\end{aligned}
\tag{2}
$$

where D_i is the ith column of D, X_p is the pth column of X, and e is a column vector of size n. Expressing the inner minimization problem as $\max_{u \in \mathbb{R}} \{u \mid u \leq$

D_i' X_p, $i = 1, \ldots, k$}, and substituting the dual of this problem back in formulation (2), produces the following bilinear program (3),

$$\min_{X,T} \sum_{i=1}^{n} D_i' X T_i$$

$$\text{s.t.} \quad e' X_p = 1 \qquad \text{for } p = 1, \ldots, k$$
$$X \in \{0, 1\}^{n \times k} \qquad\qquad\qquad (3)$$
$$\hat{e}' T_i = 1 \qquad \text{for } i = 1, \ldots, n$$
$$T \geq \mathbf{0},$$

where $T_{k \times n}$ is a matrix of variables that comes from the dual formulation. In [13], it is shown that the optimal solution to formulation (3) satisfies $\overline{X}_i \, \hat{e} \leq 1$ for $i = 1, \ldots, n$, where $\overline{X}_i$ is the ith row in the optimal solution. Therefore including this condition to formulation (3) as an additional constraint does not change the optimal solution, however helps the uncoupling of formulation (3) into two linear programming formulations.

When the values of one of the matrices X and T is fixed, then the resulting model is a linear program whose decision variables constitute the other matrix. Solving these two linear programs iteratively converges to a local optimum of the original formulation as shown in [13]. Let the fixed values $X = X^0$ satisfy the constraints of the original formulation. The resulting first linear program is given in (4),

$$T^* = LP_1(X^0):$$

$$\min_{T} \sum_{i=1}^{n} D_i' X^0 T_i \qquad\qquad\qquad (4)$$
$$\text{s.t.} \quad \hat{e}' T_i = 1 \qquad \text{for } i = 1, \ldots, n$$
$$T \geq \mathbf{0}.$$

Formulation (4) dictates that each sample is assigned to the closest median, which has an explicit solution given in expression (5),

$$T_i^* = \left\{ e(p^*) \mid p^* = \arg \min_{p=1,\ldots,k} D^0_{\,i,p} \right\} \qquad \text{for } i = 1, \ldots, n, \qquad (5)$$

where $e(p)$ is a column matrix of size k whose pth entry is 1 and the others are 0, and $D^0 = D\, X^0$ is the matrix whose entry in the ith row and pth column is the distance from point i to median p. Setting $T^0 = T^*$ in the original formulation and using the equivalence $\sum_{i=1}^{n} D_i' X T_i^0 = \sum_{p=1}^{k} \overline{T}^0_p D X_p$ in the objective function, the second linear program can be written as in (6),

$$X^* = LP_2(T^0):$$

$$\min_{X} \sum_{p=1}^{k} \overline{T}^0_p D X_p \qquad\qquad\qquad (6)$$
$$\text{s.t.} \quad e' X_p = 1 \qquad \text{for } p = 1, \ldots, k$$
$$\hat{e}' \overline{X}_i \leq 1 \qquad \text{for } i = 1, \ldots, n.$$

Algorithm 1 DKM-LP

Initialize X^0
$Z^0 \leftarrow \infty$
$t \leftarrow 1$
repeat
   Solve $T^t \leftarrow LP_1(X^{t-1})$
   Solve $X^t \leftarrow LP_2(T^t)$
   $X \leftarrow X^t$
   $t \leftarrow t + 1$
until $Z^t - Z^{t-1} < \varepsilon$
return X

It is easy to observe that formulation (6) is an assignment problem with the cost matrix $T^0 D$, hence it can be formulated as a network flow problem with X being continuous variables instead of binary variables.

The iterative uncoupled bilinear programming approach to solve discrete k-median problem is summarized in Algorithm 1, where t is the iteration counter, the input/output for the two problems are $T^t = LP_1(X^{t-1})$ and $X^t = LP_2(T^t)$ at iteration t, and Z^t is the objective function value at the end of iteration t.

Algorithm 1 converges to a local optima of the original bilinear formulation after only a few iterations [13].

To summarize in words: at each iteration, given median samples, the first linear program assigns each sample to a closest median to create clusters, and the second linear program chooses new medians based on the new clusters, which become the medians at the beginning of the next iteration.

2.1.2 A Faster Algorithm

The first linear program has an efficient explicit solution, and the second linear program is a minimum cost assignment problem, which can also be solved relatively efficiently but not as fast as the first one. Note that the selection of medians in the second linear program is not limited to the samples within their respective clusters. In almost all practical cases, the median sample of a cluster is a sample within the cluster, although it is easy to find counter examples [13]. With an additional constraint that limits the choice of a new median for a cluster among the samples in the cluster, the second linear program can also be solved explicitly. After this constraint is added, the iterative algorithm reduces to a version of Lloyd's algorithm [23], modified for discrete k-median problem.

Let M represent the indices of a set of median samples, where the jth index in M is denoted as M_j, and G_j be the set of indices of samples assigned to cluster j. The objective for a given set of median samples is calculated as,

$$Z(M) = \sum_{i=1}^{|S|} \min_{j=1,\dots,k} \{D_{i,M_j}\}, \tag{7}$$

Algorithm 2 $M = DKM(S, D, k)$

$M \leftarrow Initialize(S, D, k)$
$G_j \leftarrow \emptyset$, for $j = 1, \ldots, k$
$t \leftarrow 1$
repeat
  for $i = 1$ to $|S|$ **do**
    $j^* \leftarrow \arg \min_{j=1,\ldots,k} \{D_{i,M_j}\}$
    $G_{j^*} \leftarrow G_{j^*} \cap i$
  end for
  for $j = 1$ to k **do**
    $M_j \leftarrow \arg \min_{i \in G_j} \{\sum_{l \in G_j} D_{i,l}\}$
  end for
  $t \leftarrow t + 1$
until $Z^t - Z^{t-1} < \varepsilon$
return M

---

with respect to a given set of medians M. Let Z^t be the objective function value at iteration t and ε be the threshold. The faster method is summarized in Algorithm 2 below.

2.2 Fuzzy Discrete k-Median Clustering (FDKM)

We introduce a fuzzy discrete k-median clustering method, which is a fuzzy version of Algorithm 2, borrowing ideas from fuzzy k-means algorithm. This method is also known as fuzzy medoids algorithm [17, 27].

2.2.1 Fuzzy k-Means Clustering

The main idea behind fuzzy clustering algorithms is to let cluster membership of a sample be a continuous measure for multiple clusters rather than a discrete measure for a single cluster. The objective is to minimize the function

$$\sum_{i=1}^{n} \sum_{j=1}^{k} u_{ij}^m \|A_i - C_j\|^2, \tag{8}$$

where u_{ij} is the degree of membership of sample i to cluster j, A_i is the vector for the ith sample in the actual domain that the samples reside, C_j is a vector representing the center for cluster j in the same domain, $\| \cdot \|$ is any norm and $m > 1$. Given cluster centers C_j, $j = 1, \ldots, k$, membership can be determined by

$$u_{ik} = \frac{\|A_i - C_j\|^{-2/(m-1)}}{\sum_{p=1}^{k} \|A_i - C_p\|^{-2/(m-1)}}, \tag{9}$$

and given cluster membership values u_{ij}, new centers are determined by

$$C_j = \frac{\sum_{i=1}^{n} u_{ij}^m A_i}{\sum_{i=1}^{n} u_{ij}^m}. \tag{10}$$

The cluster memberships and cluster centers are calculated iteratively until the difference in the objective function between iterations drops below a threshold value.

2.2.2 Modification for the Discrete k-Median Clustering

We modify the fuzzy k-means algorithm following the basic idea behind the discrete k-median problem, which is the fact that arbitrary distances are used and they are not squared as in the case of k-means, and the cluster medians are selected among the samples at each iteration instead of weighted cluster centers. Given a set of median samples M, the objective is calculated as,

$$Z(M, U) = \sum_{i=1}^{n} \sum_{j=1}^{k} u_{ij}^m D_{i,M_j}, \tag{11}$$

and the cluster memberships are determined as,

$$u_{ij} = \frac{\overline{D}_{i,M_j}}{\sum_{t=1}^{k} \overline{D}_{i,M_j}}, \tag{12}$$

where $\overline{D}_{ij} = D_{ij}^{-1/(m-1)}$. Let $U_{n \times k}$ be the matrix of membership values u_{ij}. Then we can obtain a score matrix $R = D\, U$, from which we determine the jth cluster median sample as,

$$M_j = \arg \min_{i=1,\dots,|S|} \{R_{ij}\}. \tag{13}$$

If a sample is optimal for more than one cluster, it is assigned to the cluster with smaller score and removed from consideration for other clusters. Let Z^t be the objective function value at iteration t and ε be the threshold. Then, we can summarize the fuzzy discrete k-median approach as shown in Algorithm 3.

2.3   Hierarchical Cluster Tree (HCT)

HCT algorithm is an agglomerative method that iteratively merges smaller clusters into larger ones until all samples are collected under a single large cluster [32]. In the process of linking these clusters, a tree structure is formed, which is referred to as

---

Algorithm 3 $M = FDKM(S, D, k)$

$M \leftarrow Initialize(S, D, k)$
$t \leftarrow 1$
repeat
    $I = \{1, \ldots, |S|\}$
    $K = \{1, \ldots, k\}$
    $U \leftarrow \{u_{i,j} : u_{i,j} = \overline{D}_{i,M_j} / \sum_{t=1}^{k} \overline{D}_{i,M_j}\}$
    $R \leftarrow D\,U$
    while $K \neq \emptyset$ **do**
        $(i^*, j^*) \leftarrow \arg\min_{i \in I}\{\min_{j \in K}\{R_{ij}\}\}$
        $M_j^* \leftarrow i^*$
        $I = I \setminus \{i^*\}$
        $K = K \setminus \{j^*\}$
    end while
    $t \leftarrow t + 1$
until $Z^t - Z^{t-1} < \varepsilon$
return M

---

a *dendrogram*. The merging process is steered with respect to a distance measure between subclusters. Once the dendrogram is completed, individual clusters can be obtained by different ways of cutting the dendrogram into sub-trees.

HCT algorithm starts with every sample defined as a cluster by itself. The distance measure that guides the linking process can be defined in a number of ways. Two most well known distance measures are *single linkage* and *complete linkage* distance measures. Let G_i and G_j be two clusters. Single linkage distance is defined as the minimum distance between any pair of samples such that one selected from G_i and the other selected from G_j. More formally,

$$\Delta_{ij} = \min\{D_{pq} : p \in G_i, q \in G_j\}. \tag{14}$$

Complete linkage distance is defined as the maximum distance between any pair of samples such that one selected from G_i and the other selected from G_j, that is,

$$\Delta_{ij} = \max\{D_{pq} : p \in G_i, q \in G_j\}. \tag{15}$$

There are other distance measures such as *average* or *median* distance, which is the average or the median of the distances between any pairs of samples. Other distance measures may involve *centroid*, *squared* or *weighted* distances. However, these methods may require operations in the actual domain of the data, which may not be practical or meaningful for time series data. Single and complete linkage distances do not require such calculations and work with an arbitrary distance matrix. In our study we use the complete linkage distance, which is also known as the *diameter* of the new cluster.

HCT starts with a distance matrix $\Delta = D$, where all samples are represented as individual clusters. At iteration t, clusters i and j with minimum Δ_{ij} among all pairs is selected and merged into a larger cluster. Matrix Δ is updated by deleting the

---

Algorithm 4 $M = HCT(S, D, k)$

$I = \{1, \ldots, |S|\}$
$\Delta \leftarrow D$
for $p = 1$ to $|S| - 1$ **do**
　　$(i^*, j^*) \leftarrow \arg\min\{\Delta_{i,j} : i, j \in I, i \neq j\}$
　　$L(p) \leftarrow (i^*, j^*, \Delta_{i^*,j^*})$
　　$G_{|S|+p} \leftarrow G_{i^*} \cup G_{j^*}$
　　$I \leftarrow I \cup \{|S| + p\}$
　　$I \leftarrow I \setminus \{i^*, j^*\}$
　　Calculate $\Delta_{i,|S|+p}$ for all $i \in I$
end for
$H \leftarrow Cluster(L)$
for $j = 1$ to k **do**
　　$M_j \leftarrow \arg\min_{q \in H_j}\{\sum_{p \in H_j} D_{q,p}\}$
end for
return M

---

rows and columns corresponding to clusters i and j, assigning a new index, $n + t$ to the merged cluster, and appending a row and column to Δ for the new cluster with the values $\Delta_{i,n+k}$ for all remaining clusters. Meanwhile, at each iteration, a node for the new cluster is created and connected to the clusters that are merged to produce the new cluster in order to continue building the dendrogram. Algorithm continues until Δ reduces down to a single element. There are exactly $m - 1$ internal nodes in the dendrogram, which can be represented by a list L of such nodes. In this list, the indices of the nodes corresponding to the two subclusters that form a new cluster, and the distances between them are stored for the new internal node representing the new cluster.

There are many ways of cutting the dendrogram into clusters. Usually, some measure is used together with a threshold, below which clusters are formed. One can calculate *inconsistency* coefficient of a link by comparing the height of a link with the average height of links at the same level in the hierarchical dendrogram. The higher the inconsistency of a link, the less similar clusters connected by that link. The leaves of the tree with inconsistency below a threshold value are grouped into a cluster. Height of a node for a cluster, which is defined as the distance between the two subclusters that form that cluster, can also be used with a threshold value to form clusters. In order to obtain a predetermined number of k clusters, one can also find the minimum height at which a horizontal cut through the dendrogram forms k clusters, which is the cutting method used to form clusters in this study. Once the clusters are formed after cutting the dendrogram, the median sample in each cluster is identified to represent these clusters.

HCT method is summarized in Algorithm 4, in which I is the set of indices for subclusters, and Δ_I is the distance matrix whose entries are based on the distance given in (15). Forming a set of clusters H from the list L by a horizontal cut is represented by the method *Cluster*, details of which are skipped for brevity.

Algorithm 5 $M = PAM(S, D, k)$

$M \leftarrow Initialize(S, D, k)$
$t \leftarrow 1$
repeat
  $N \leftarrow S \setminus M$
  $Z_{min} \leftarrow \infty$
  $M_{min} \leftarrow M$
  for each $i \in N, j \in M$ **do**
    $M^* \leftarrow (M \cup N_i) \setminus M_j$
    if $Z(M^*) < Z_{min}$ **then**
      $Z_{min} \leftarrow Z(M^*)$
      $M_{min} \leftarrow M^*$
    end if
  end for
  $M \leftarrow M_{min}$
  $t \leftarrow t + 1$
until $Z^t - Z^{t-1} < \varepsilon$
return M

2.4 Partition Around Medoids (PAM)

PAM algorithm is a simple heuristic based on updating a set of medians, which are also referred to as *medoids*, by exhaustively swapping medoid samples with non-medoid samples to find the best possible improvement at each iteration. Given a set of medoids, each sample is assigned to the nearest medoid, and the sum of all distances between samples and their assigned medoids is calculated as the cost of the current configuration using expression (7). At each iteration, each medoid sample is temporarily swapped with each non-medoid sample and the cost of the new arrangement is calculated. The swap that produces the minimum cost becomes permanent at the end of the iteration. The algorithm continues until the reduction in the total cost between two iterations falls below a threshold value. PAM method is summarized in Algorithm 5.

2.5 Initialization

All methods introduced, except for HCT, are iterative methods starting with a set of median samples and converging to a local optimum when their threshold values are set to be zero. Clustering problem is a hard combinatorial problem with a large number of local optima. Therefore, the solutions obtained from these algorithms are quite sensitive to the initial selection of medians.

For DKM, FDKM and PAM methods, we use the initialization method originally included in the PAM algorithm [19] to build a "good" set of initial medians. We start building the initial set by including the sample whose sum of distances to other

---

Algorithm 6 $M = Initialize(S, D, k)$

$M_1 \leftarrow \arg\min_{i \in S} \sum_{j \in S} D_{i,j}$
for $t = 2$ to k **do**
  $N \leftarrow S \setminus M$
  for each $p \in N$ **do**
    $D_{i,M} \leftarrow \min\{D_{i,j} : j \in M\}$ for all $i : i \in N, i \neq p$
    $T_p = \sum_{i:i \neq j, i \in N} \max\{D_{i,M} - D_{i,p}, 0\}$
  end for
  $p^* \leftarrow \arg\max_{p \in N}\{T_p\}$
  $M \leftarrow M \cup \{p^*\}$
end for
return M

---

samples is minimum. Then we determine the remaining $k - 1$ initial medians as follows: Let M be the current set of median samples and $N = S \setminus M$ be the set of remaining samples. Let the smallest distance of a sample to the set of medians be denoted as $D_{i,M} = \min\{D_{i,j} : j \in M\}$. For a new candidate median sample p, let the total positive distance gain T_p be measured as the sum of the positive difference between the minimum distance of every nonmedian samples $i : i \neq j$, $i \in N$ to the current median samples and to the candidate sample p, that is, $T_p = \sum_{i:i \neq j, i \in N} \max\{D_{i,M} - D_{i,p}, 0\}$. At each of the remaining $k - 1$ iterations we remove the sample with the maximum total distance gain from N and add it to M. The initialization method is given in Algorithm 6.

3 Computational Results

We code the four clustering algorithms introduced, namely, DKM, FDKM, HCT and PAM in MATLAB 7.9 and apply them to the University of California at Riverside (UCR) benchmark time series data sets, and to real life local field potential (LFP) recordings collected from the occipital lobe of a macaque monkey brain during a visuomotor task. We use *normalized mutual information*(NMI) measure [29] on UCR benchmark data [35], LFP data [10], and we graphically demonstrate NMI based discrimination of visual stimuli from the LFP recordings. We also compare running time results from UCR datasets.

A distance or dissimilarity matrix is sufficient for all of the four methods without the actual knowledge of domain that the samples reside in. We use four different distance measures: rectilinear, Euclidean, squared-Euclidean, and DTW. Among these distance measures, squared-Euclidean is the only Bregman divergence, a class of non-metric distance measures that are commonly used in clustering. Using squared-Euclidean distances is equivalent to k-means algorithms [3,9]. Therefore, we note that the results for the squared-Euclidean can be considered as results of a discrete k-means algorithm in geometric space.

Since DKM, FDKM and PAM methods are sensitive to the initial set of medians, we adopt the initialization procedure of the PAM algorithm to create a set of good starting medians. We set the threshold $\varepsilon = 0$ for all computational results to terminate all of the three methods at a local optimum.

3.1   Distance Measures

We summarize the four distance measures used to create the distance matrix D. Assume S_i is a multi-variate time series sample which is composed of d single-variate time series with n time points each. S_i is represented with a $d \times n$ matrix. The values at time point t is denoted with the vector $S_i(t)$.

3.1.1   Rectilinear

This distance measure is the sum of the absolute values of the differences between all corresponding elements of the two multi-variate time series samples S_i and S_j, and is calculated as,

$$D_{rtl}(i, j) = \sum_{t=1}^{n} ||S_i, (t) - S_j(t)||_1. \tag{16}$$

3.1.2   Euclidean

This distance measure is the sum of the Euclidean norm of the difference between the corresponding vectors $S_i(t)$ and $S_j(t)$ over each time point t, which is given as,

$$D_{euc}(i, j) = \sum_{t=1}^{n} ||S_i(t) - S_j(t)||_2. \tag{17}$$

3.1.3   Squared-Euclidean

This distance measure is similar to Euclidean, where the square of the norm of the difference vector between $S_i(t)$ and $S_j(t)$ are summed over time t, that is,

$$D_{euc^2}(i, j) = \sum_{t=1}^{n} ||S_i(t) - S_j(t)||_2^2 \tag{18}$$

Algorithm 7 $d = D_{dtw}(i, j)$

$C_1 \leftarrow ||S_i(1) - S_j(1)||_2$
for $p = 2$ to n **do**
$\quad C_p \leftarrow C_{p-1} + ||S_i(p) - S_j(1)||_2$
end for
for $q = 2$ to n **do**
$\quad d \leftarrow C_1 + ||S_i(1) - S_j(q)||_2$
$\quad$ **for** $p = 2$ to n **do**
$\quad\quad temp \leftarrow \min\{up, C_{p-1}, Cp\} + ||S_i(p) - S_j(q)||_2$
$\quad\quad C_{p-1} \leftarrow d$
$\quad\quad d \leftarrow temp$
$\quad$ **end for**
$\quad C_n \leftarrow c$
end for
return d

3.1.4 Dynamic Time Warping (DTW)

This distance measure calculates the similarities of two time series by deleting, inserting or matching components of one time series compared to another time series. The optimal set of editing operations are determined by dynamic programming. The cumulative cost of the time series is calculated as the sum of the Euclidean norm of the distances between the vectors of the matched time points between the two series. Although in practice DTW usually employs an $n \times n$ matrix to identify the actual matching between the two time series, here, we only need the cumulative matching distance, which only requires a vector $C_{n \times 1}$, which is updated n times. The method used to calculate the cumulative DTW distance is summarized in Algorithm 7.

3.1.5 Kernel Induced Spaces

Although kernel induced spaces are not included in the computational results, we note that it is straightforward to create a distance matrix for such spaces. The actual domain S for the original data can be mapped to a usually higher dimensional feature space using a mapping $\Phi(S)$. A kernel function $K(S_i, S_j) = \langle \Phi(S_i) \cdot \Phi(S_j) \rangle$, which is generally in the form of a dot product, implicitly maps the data without creating the actual representation of the samples in the feature space. The kernel function can be used to measure the squared-Euclidean distance in the kernel space, that is,

$$||\Phi(S_i) - \Phi(S_j)||_2^2 = \langle \Phi(S_i), \Phi(S_i) \rangle + \langle \Phi(S_j), \Phi(S_j) \rangle - 2 \langle \Phi(S_i), \Phi(S_j) \rangle$$

$$= K(S_i, S_i) + K(S_j, S_j) - 2 K(S_i, S_j). \tag{19}$$

The square root of expression (19) is the actual Euclidean distance in the feature space, or it can be used as is for the squared distances in the feature space.

3.2 Performance Measure

For clustering performance, we use *normalized mutual information* (NMI) [29], which measures the normalized mutual dependence of two variables. Its calculation is based on entropies of two sets of labels and the mutual information in between them. Let $G = \{G_i\}_{i=1,...,k}$ be the actual clusters in S and $G^* = \{G_j^*\}_{j=1,...,k}$ be the clusters obtained from S using a clustering method. Then, the mutual information I between G^* and G is defined as,

$$I(G^*, G) = \sum_{i=1}^{k} \sum_{j=1}^{k} \frac{|G_i^* \cap G_j|}{|S|} \log \frac{|S| \, |G_i^* \cap G_j|}{|G_i^*| \, |G_j|}. \tag{20}$$

The entropy of a set is defined as,

$$H(G) = -\sum_{i=1}^{k} \frac{|G_i|}{|S|} \log \frac{|G_i|}{|S|}. \tag{21}$$

Then NMI between G^* and G can be expressed as,

$$NMI(G^*, G) = \frac{I(G^*, G)}{[H(G^*) + H(G)]/2}. \tag{22}$$

3.3 UCR Benchmark Datasets

We use the University of California - Riverside (UCR) time series data sets [35] to measure clustering performances of DKM, FDKM, HCT and PAM methods. NMI scores from these methods are compared using four different distance measures, namely, rectilinear, Euclidean, squared-Euclidean, and DTW. UCR data base consists of 20 real world and synthetic time series data sets with various time series data properties. Characteristics of the data sets can be found in Table 1, including number of classes, data size and temporal length. UCR data sets are generally used for classification, and therefore, they are split into training and testing sets. We apply the four clustering methods on the training data sets to identify median samples and cluster both training and testing data sets using these medians. We report NMI values for both training and testing data sets in Tables 2 and 3, respectively.

For each dataset, the highest NMI value is emphasized in bold. The background for the distance measure with the highest NMI result within each clustering method for each data set is also colored gray. In both the training and testing results, it is visible that DTW is overwhelmingly superior to other distance measures, especially in the training data sets. In the testing data sets, Euclidean and squared-Euclidean measures have second and third highest numbers, followed by rectilinear distances.

Table 1 University of
California-Riverside time
series data sets by [35]

Dataset	Name	Classes	Size	Length
1	50 word	50	450	270
2	Adiac	37	390	176
3	Beef	5	30	470
4	CBF	3	30	128
5	Coffee	2	38	286
6	ECG200	2	100	96
7	FaceAll	14	510	131
8	FaceFour	4	24	350
9	Fish	7	175	463
10	Gun Point	2	50	150
11	Lightning 2	2	60	637
12	Lightning7	7	70	319
13	Olive Oil	4	30	570
14	OSU Leaf	6	200	427
15	Swedish Leaf	15	500	128
16	Synthetic Control	6	300	60
17	Trace	4	1,000	275
18	Two Patterns	4	1,000	128
19	Wafer	2	1,000	152
20	Yoga	2	300	426

For the data sets with NMI scores more than 50%, PAM leads in terms of the highest NMI scores, followed by DKM, FDKM, and HCT both in training and testing results. Datasets with lower NMI scores have similar results in most cases.

For each clustering method, running time does not change much for different distance measures, however there is a significant difference in running times among these methods. In Table 4, running times from the UCR data sets are presented for each clustering method as the average of the running times for the four different distance measures. The actual running times are given in the first half of the table in milliseconds and normalized running times based on DKM, which is the overall fastest method, is given in the second half of the table. The results suggest that FDKM is slightly slower than DKM, HCT is four times as slow in most data sets, and PAM is several orders of magnitude slower than DKM. These results, combined with the clustering performance, show that DKM and FDKM methods are both efficient and effective in clustering time series data using a distance matrix.

3.4   Local Field Potentials from a Visuomotor Task

The neural data we study is the local field potentials (LFP) collected from multiple channels implanted in different cortical areas of a macaque monkey during a visual discrimination task [10]. We include LFP recordings from three electrodes in the occipital lobe, which is the visual processing center of the brain. Experiments

Table 2 NMI scores for UCR training data sets with respect to DKM, FDKM, HCT and PAM methods using rectilinear (RTL), Euclidean (EUC), squared-Euclidean (EUC^2), and dynamic time warping (DTW), distance measures. For each dataset, best scores within a method are shown with *gray* background and overall best scores are shown in *bold*

	DKM				FDKM			
Dataset	RTL	EUC	EUC^2	DTW	RTL	EUC	EUC^2	DTW
1	0.6797	0.6552	0.6622	0.6976	0.6803	0.6521	0.6577	0.6966
2	0.6186	0.6222	0.6044	0.6406	0.6184	0.6179	0.6071	**0.6413**
3	0.4188	0.4188	0.4800	0.4188	0.4188	0.4800	0.4800	0.4188
4	0.2687	0.2698	0.2698	**0.8955**	0.2687	0.2698	0.2698	0.7370
5	0.0357	0.0044	0.0044	**0.1565**	0.0000	0.0000	0.0615	0.1009
6	0.1402	**0.1751**	0.1551	0.1589	0.1551	0.1551	**0.1751**	0.1589
7	0.4298	0.3858	0.3897	0.7613	0.4352	0.3858	0.3849	0.7422
8	0.5364	0.3632	0.5442	**0.7851**	0.5364	0.3979	0.5442	**0.7851**
9	0.3520	0.3827	0.3636	0.4249	0.3520	0.3845	0.3884	0.4249
10	0.0126	0.0126	0.0126	0.0227	0.0126	0.0126	0.0126	0.0227
11	0.0151	0.0219	0.0263	0.0673	0.0206	0.0600	0.0219	0.0973
12	0.5165	0.4435	0.4393	0.5479	0.5182	0.4435	0.4501	0.5390
13	0.5803	0.5827	**0.6723**	0.4967	0.5803	0.5827	**0.6723**	0.4967
14	0.1873	0.1879	0.2054	0.2720	0.1873	0.1780	0.2087	0.2720
15	0.5355	0.5737	0.5147	0.5614	0.5391	0.5719	0.5296	0.5612
16	0.5635	0.5738	0.5730	0.9090	0.5554	0.5761	0.5667	0.9036
17	0.5679	0.5113	0.5113	**0.7442**	0.5718	0.5069	0.5113	**0.7442**
18	0.0186	0.0195	0.0195	0.9122	0.0265	0.0195	0.0195	0.9035
19	0.0002	0.0000	0.0000	0.0004	0.0002	0.0001	0.0000	**0.0011**
20	**0.0051**	**0.0051**	0.0030	0.0012	**0.0051**	**0.0051**	0.0030	0.0012

	HCT				PAM			
Dataset	RTL	EUC	EUC^2	DTW	RTL	EUC	EUC^2	DTW
1	0.6687	0.6708	0.6708	**0.7192**	0.6803	0.6713	0.6685	0.7123
2	0.5518	0.5779	0.5779	0.6198	0.5924	0.6046	0.5855	0.6278
3	0.4679	0.4294	0.4294	0.4558	0.4188	0.4800	**0.4908**	0.4188
4	0.4619	0.3690	0.3690	0.6741	0.2687	0.2698	0.2698	**0.8955**
5	**0.1565**	**0.1565**	**0.1565**	**0.1565**	0.0597	0.0948	0.0948	0.0948
6	0.1402	0.1709	0.1709	**0.1751**	0.1402	**0.1751**	0.1594	0.1589
7	0.3533	0.3492	0.3492	0.6601	0.4124	0.3941	0.4047	**0.7891**
8	0.5196	0.4729	0.4729	0.6751	0.6031	0.4839	0.4839	**0.7851**
9	0.2700	0.2461	0.2461	**0.4349**	0.3769	0.3879	0.2764	0.4105
10	0.0412	0.0412	0.0412	**0.1386**	0.0126	0.0126	0.0126	0.0126
11	0.0108	0.0388	0.0388	**0.2676**	0.0151	0.0219	0.0263	0.1090
12	0.5520	0.3810	0.3810	**0.6223**	0.5165	0.4599	0.4495	0.5920
13	0.4951	0.5000	0.5000	0.5000	0.5803	0.5827	**0.6723**	0.4967
14	0.0883	0.2337	0.2337	0.2856	0.1934	0.2106	0.2162	**0.2995**
15	0.4185	0.3632	0.3632	0.4972	0.5346	0.5407	0.5359	**0.5961**
16	0.6104	0.5627	0.5627	0.8137	0.5726	0.5702	0.5718	**0.9439**
17	0.5729	0.5729	0.5729	0.5270	0.5088	0.5113	0.5545	0.7417
18	0.0364	0.0185	0.0185	0.4399	0.0232	0.0259	0.0259	**0.9218**
19	0.0000	0.0001	0.0001	0.0000	0.0002	0.0000	0.0000	0.0004
20	0.0025	0.0001	0.0001	0.0002	**0.0051**	**0.0051**	0.0030	0.0012

Table 3 NMI scores for UCR testing data sets with respect to DKM, FDKM, HCT and PAM methods using rectilinear (RTL), Euclidean (EUC), squared-Euclidean (EUC^2), and dynamic time warping (DTW), distance measures. For each dataset, best scores within a method are shown with *gray* background and overall best scores are shown in **bold**

Dataset	DKM				FDKM			
	RTL	EUC	EUC^2	DTW	RTL	EUC	EUC^2	DTW
1	0.6520	0.6426	0.6427	0.7153	0.6545	0.6398	0.6420	0.7152
2	0.5896	0.6181	0.5935	0.6121	0.5863	**0.6202**	0.5930	0.6105
3	0.2856	0.3454	0.3251	0.3220	0.2856	0.3565	0.3565	0.3518
4	0.2896	0.3235	0.3235	**0.7194**	0.2896	0.3235	0.3235	0.6440
5	0.0381	0.0459	0.0459	0.0308	0.0252	0.0100	0.0381	**0.0916**
6	0.1194	**0.1506**	0.1344	0.0644	0.1344	**0.1506**	**0.1506**	0.0644
7	0.3197	0.2811	0.2707	0.6190	0.3194	0.2811	0.2735	0.5843
8	0.3269	0.2626	0.2795	**0.6172**	0.3269	0.2461	0.2629	**0.6172**
9	0.3467	0.3269	0.3203	0.4177	0.3467	0.3570	0.3543	0.4177
10	0.0011	0.0011	0.0011	0.0011	0.0011	0.0011	0.0011	0.0011
11	0.0083	0.0568	0.1231	0.0209	0.0097	0.0418	0.0348	0.0121
12	0.5288	0.3810	0.3756	0.5306	0.5419	0.3810	0.3945	0.5062
13	0.2519	0.4487	**0.4714**	0.3047	0.2519	0.4487	**0.4714**	0.3047
14	0.1677	0.1781	0.1729	0.2369	0.1677	0.1694	0.1771	0.2369
15	0.5476	0.5796	0.5279	0.5686	0.5459	0.5722	0.5421	0.5685
16	0.5663	0.5528	0.5827	0.8857	0.5376	0.5485	0.5769	0.8833
17	0.5410	0.5109	0.5099	0.7693	0.5329	0.5090	0.5109	**0.7769**
18	0.0222	0.0244	0.0244	0.9015	0.0325	0.0244	0.0244	0.8994
19	0.0000	0.0004	0.0004	0.0000	0.0000	**0.0005**	0.0004	0.0001
20	0.0011	**0.0013**	0.0003	0.0006	0.0011	**0.0013**	0.0003	0.0006

Dataset	HCT				PAM			
	RTL	EUC	EUC^2	DTW	RTL	EUC	EUC^2	DTW
1	0.6460	0.6433	0.6431	0.6996	0.6699	0.6616	0.6501	**0.7170**
2	0.5705	0.5564	0.5577	0.5516	0.5821	0.6128	0.5836	0.6080
3	0.3500	0.3437	**0.3842**	0.3163	0.2856	0.3565	0.3565	0.2898
4	0.2471	0.2937	0.2937	0.6060	0.2896	0.3235	0.3235	0.6850
5	0.0308	0.0308	0.0308	0.0308	0.0336	0.0336	0.0336	0.0336
6	0.1194	0.1349	0.1349	0.0925	0.1194	**0.1506**	0.1349	0.0644
7	0.2779	0.2839	0.2790	0.5340	0.3176	0.2833	0.2911	**0.6754**
8	0.5265	0.4781	0.4870	0.4529	0.3856	0.4050	0.4050	**0.6172**
9	0.2451	0.2594	0.2444	**0.4355**	0.3021	0.3159	0.2745	0.4043
10	0.0779	0.0687	0.0687	**0.1237**	0.0011	0.0011	0.0011	0.0011
11	0.0064	0.1231	0.1231	0.0047	0.0083	0.0568	0.1231	**0.1381**
12	**0.6074**	0.4537	0.4361	0.5583	0.5288	0.3591	0.3611	0.5812
13	0.3580	**0.4714**	**0.4714**	0.4111	0.2519	0.4487	**0.4714**	0.3047
14	0.1532	0.1763	0.1747	0.2176	0.1867	0.1888	0.1680	**0.2638**
15	0.3742	0.4010	0.3936	0.4930	0.5561	0.5669	0.5616	**0.6178**
16	0.5556	0.5610	0.5378	0.7776	0.5775	0.5805	0.5866	**0.8993**
17	0.5347	0.5502	0.5481	0.5461	0.5078	0.5109	0.5398	0.7736
18	0.0395	0.0265	0.0304	0.3715	0.0277	0.0264	0.0264	**0.9053**
19	0.0000	0.0004	0.0004	0.0000	0.0000	0.0004	0.0004	0.0000
20	0.0003	0.0011	0.0010	0.0000	0.0011	**0.0013**	0.0003	0.0006

Table 4 UCI data sets training time results: Actual running time is the average of running times over different distance measures which have similar magnitude within same data set. Normalized time results are reported using DKM running time as the basis

Dataset	Actual speed (milliseconds)				Normalized speed			
	DKM	FDKM	HCT	PAM	DKM	FDKM	HCT	PAM
1	54.27	70.93	145.03	121,367.88	1.00	1.31	2.67	2,236.24
2	2.59	8.64	17.68	45,339.59	1.00	3.33	6.82	17,488.46
3	0.40	0.54	1.28	62.42	1.00	1.33	3.19	155.38
4	0.21	0.22	0.80	22.17	1.00	1.06	3.90	107.95
5	0.21	0.23	0.75	28.39	1.00	1.09	3.53	134.39
6	0.25	0.23	1.32	64.35	1.00	0.93	5.37	262.62
7	1.21	5.48	42.40	15,018.71	1.00	4.53	35.08	12,425.78
8	0.24	0.34	0.99	38.53	1.00	1.40	4.05	157.56
9	0.43	0.91	2.83	1,068.96	1.00	2.13	6.63	2,503.67
10	0.29	0.40	1.24	59.65	1.00	1.37	4.25	204.54
11	0.30	1.74	1.36	53.61	1.00	5.87	4.58	180.66
12	0.30	0.47	1.20	290.04	1.00	1.58	4.03	972.45
13	0.16	0.27	0.83	21.28	1.00	1.69	5.31	135.26
14	0.44	0.77	3.42	1,150.84	1.00	1.76	7.82	2,629.21
15	1.63	7.46	40.67	14,851.99	1.00	4.57	24.89	9,089.31
16	0.72	1.00	11.57	2,234.17	1.00	1.38	16.01	3,092.55
17	0.49	0.60	1.52	342.84	1.00	1.22	3.10	698.79
18	5.07	13.54	110.43	5,144.03	1.00	2.67	21.80	1,015.23
19	13.51	10.38	117.30	1,196.61	1.00	0.77	8.68	88.57
20	1.11	0.81	10.20	278.31	1.00	0.74	9.22	251.71

involve repeated trials of one of the four visual stimuli: right slanted line, right slanted diamond, left slanted line, and left slanted diamond. Visual stimulus is shown to the monkey on a screen in a randomized order. Each type of stimulus is shown about a quarter of 200 trials per session in a randomized order. We aim to use clustering methods to identify the time, at which presentation of different stimuli in the occipital lobe in terms of local field potentials (LPF) creates different clusters. We use a window of 10 recordings from the three occipital lobe channels, which is equivalent to a window of 50 ms. We slide this window from 100 ms prior to the onset of the stimuli until 500 ms after the stimuli onset, advancing 5 points, or 25 ms at a time, and applying the four clustering methods introduced for every window.

In Fig. 1, results of the clustering efforts are shown in terms of NMI. In each graph, the NMI scores for the four different clustering methods are plotted over time with the distance measure used indicated below the graph. Euclidean and squared-Euclidean distance measures produce very similar results, better than rectilinear but not with an as steep increase as DTW at the beginning. It is worth noting that rectilinear distances produce the highest peak but least steep graph. DKM and FDKM produce very similar results with FDKM having a slight advantage around the peak. Results from HCT are dominated by the other three in each graph, and PAM has a slight advantage over DKM and FDKM.

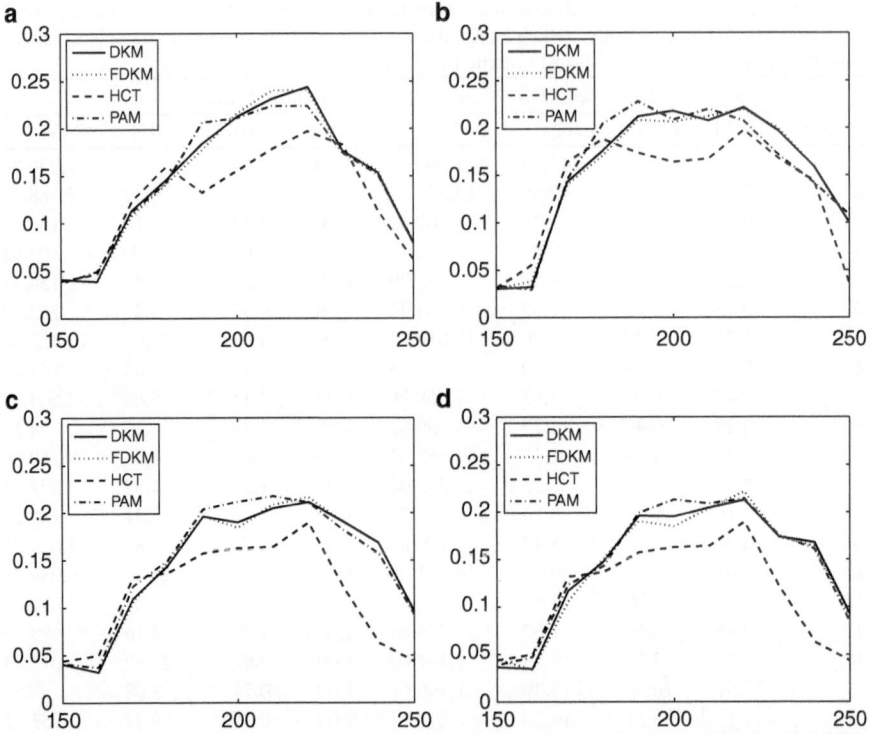

Fig. 1 NMI scores for LFP data sets between 100 and 200 ms with respect to DKM, FDKM, HCT and PAM methods using (**a**): Rectilinear, (**b**): Dynamic time warping, (**c**): Euclidean, and (**d**) squared-Euclidean distance measures

In each graph, change starts around 170 ms and peaks around 200 ms, which is 70 and 100 ms after the onset of the visual stimuli, respectively. The results shown in the graphs come from one of the ten sessions processed. Focusing on the time window that correspond to the peak at 100 ms after the onset of the stimuli, we apply the four clustering methods together using four different distance measures to each session separately. The results are presented in Table 5. As in UCR results, bold numbers are the best across clustering methods and distance measures for each session. FDKM has the most number of highest NMI scores followed by DKM, PAM and HCT in decreasing number of highest NMI scores. The NMI scores with gray background are the highest NMI scores among different distance measures within a clustering algorithm for each session. DTW seems to work well with all clustering algorithms, followed by rectilinear, Euclidean, and squared-Euclidean.

Table 5 NMI scores for LFP data sets with respect to DKM, FDKM, HCT and PAM methods using rectilinear (RTL), Euclidean (EUC), squared-Euclidean (EUC^2), and dynamic time warping (DTW), distance measures. For each dataset, best scores within a method are shown with *gray* background and overall best scores are shown in **bold**

Dataset	DKM				FDKM			
	RTL	EUC	EUC^2	DTW	RTL	EUC	EUC^2	DTW
1	0.0925	0.0944	**0.1008**	0.0828	0.0925	0.0921	0.0939	0.0814
2	0.0891	0.0958	0.0943	0.1024	**0.1056**	0.0896	0.0926	0.1024
3	0.0629	0.0590	0.0574	0.0645	0.0677	0.0571	0.0631	0.0672
4	0.1201	0.1108	0.1198	0.1215	**0.1332**	0.1100	0.1174	0.1204
5	0.1110	0.1248	0.1133	0.1126	0.1113	**0.1351**	0.1218	0.1130
6	0.2076	0.2090	0.2114	0.2003	0.1964	0.2027	0.1975	0.1975
7	0.2068	0.1995	0.2050	0.2313	0.2216	0.2160	0.2201	**0.2505**
8	**0.2439**	0.2115	0.2131	0.2212	0.2409	0.2168	0.2217	0.2194
9	0.1750	0.1778	0.1798	0.2017	0.1721	0.1774	0.1760	0.1831
10	0.2047	0.2091	0.2052	0.2066	0.1994	0.2033	0.1995	0.2073

Dataset	HCT				PAM			
	RTL	EUC	EUC^2	DTW	RTL	EUC	EUC^2	DTW
1	0.0762	0.0897	0.0897	0.0887	0.0911	0.0831	0.0849	0.0821
2	0.0944	0.0972	0.0972	0.0854	0.1029	0.0986	0.0960	0.1009
3	0.0667	0.0630	0.0630	0.0594	**0.0681**	0.0615	0.0606	0.0668
4	0.1028	0.1085	0.1085	0.1108	0.1195	0.1283	0.1237	0.1319
5	0.1191	0.1027	0.1027	0.1141	0.0987	0.1049	0.1099	0.1158
6	0.1601	0.1718	0.1718	0.1827	0.2052	0.1982	0.1963	0.2058
7	0.1808	0.1791	0.1791	0.1946	0.2077	0.2193	0.1962	0.2093
8	0.1970	0.1895	0.1895	0.1975	0.2238	0.2177	0.2143	0.2277
9	0.1698	**0.2028**	**0.2028**	0.1795	0.1681	0.1837	0.1859	0.1893
10	0.1822	0.1977	0.1977	0.1954	**0.2276**	0.2131	0.2166	0.2226

4 Conclusion

In this study, we introduce four clustering methods, which take a distance matrix that represents dissimilarities between samples as input. Three of these methods are partitional methods and the fourth method is a hierarchical method. Dependence on a distance matrix spares the difficulties or complications of operating in the actual domain that the data resides, and provides more flexibility in the way the dissimilarities between samples are defined. All four methods produce k samples as medians that represent clusters such that samples are assigned to the median closest to them and form clusters.

The first partitional method is based on the transformation of the exact clustering problem into a bilinear program with uncoupled variables, which can be solved iteratively as two alternating linear programs until it converges to local optimum. One of these linear programs has an explicit solution, whereas the other is a minimum cost assignment problem. With an additional constraint on the assignment problem, which hardly affects the optimal solution in most cases, the uncoupled bilinear approach is converted into a discrete k-median (DKM) method, which is the discrete version of Lloyd's algorithm. A fuzzy version of the discrete k-median algorithm (FDKM) is included as the second clustering method, which uses fuzzy membership degrees to all clusters for each sample. These membership values are iteratively updated and new cluster medians are found at each iteration until the algorithm converges to a local optimum. Hierarchical cluster tree method (HCT) with complete linkage is presented as the third method, which builds an agglomerative tree called dendrogram. This tree is cut horizontally to form k clusters and the median sample in each cluster is returned. Partition around medoids (PAM), an iterative method that swaps median and non-median samples to improve clustering is introduced as the fourth clustering method.

Distance measures are selected as rectilinear, Euclidean, squared-Euclidean and dynamic time warping (DTW) to form the distance matrices from single or multivariate time series. All combinations of the four clustering methods and four distance measures are applied to the University of California at Riverside (UCR) data sets and local field potential (LFP) recordings from the occipital lobe of a macaque monkey. The UCR results suggest that the DTW distance measures work better than the other distance measures. PAM produces slightly better results in UCR data sets, closely followed by DKM and FDKM, but PAM is several orders of magnitude slower than DKM and FDKM clustering algorithms. HCT performs well in some data sets but is also few times slower than DKM and FDKM. In the LFP results, distance measures do not make a big impact, however DKM and FDKM results are better than PAM and HCT in most data sets. All four methods are capable of detecting the time of the discrimination of visual stimulus in the occipital lobe.

References

1. K.P. Bennett, O.L. Mangasarian, Bilinear separation of two sets in n-space. Comput. Optim. Appl. **2**, 207–227 (1993)
2. D.J. Berndt, J. Clifford, in *Using Dynamic Time Warping to Find Patterns in Time Series*. Proceedings of KDD-94: AAAI Workshop on Knowledge Discovery in Databases, Seattle, Washington, pp. 359–370 (1994)
3. J. Blömer, M.R. Ackermann, C. Sohler, Clustering for metric and nonmetric distance measures. ACM Trans. Algorithms **6**, 59:1–59:26 (2010)
4. P.S. Bradley, U.M. Fayyad, in *Refining Initial Points for k-Means Clustering*. ICML '98: Proceedings of the Fifteenth International Conference on Machine Learning, San Francisco, CA, USA, 1998 (Morgan Kaufmann, CA, 1998), pp. 91–99
5. J.F. Campbell, Integer programming formulations of discrete hub location problems. Eur. J. Oper. Res. **72**, 387–405 (1994)
6. J.F. Campbell, Hub location and the p-hub median problem. Oper. Res. **44**, 923–935 (1996)
7. D. Chhajed, T.J. Lowe, m-median and m-center problems with mutual communication: Solvable special cases. Oper. Res. **40**, S56–S66 (1992)
8. P. Chuchart, S. Supot, C. Thanapong and S. Manas. Automatic segmentation of blood vessels in retinal image based on fuzzy k-median clustering. In Proceedings of the 2007 IEEE International Conference on Integration Technology, pp. 584–588, 2007.

9. I.S. Dhillon, A. Banerjee, S. Merugu, J. Ghosh, Clustering with bregman divergences. J. Mach. Learn. Res. **6**, 1705–1749 (2005)
10. M. Ding, R. Coppola, A. Ledberg, S.L. Bressler, R. Nakamura, Large-Scale Visuomotor Integration in the Cerebral Cortex. Cerebr. Cortex **17**(1), 44–62 (2007)
11. P. D'Urso, R. Coppi, P. Giordani, in *Fuzzy k-Medoids Clustering Models for Fuzzy Multivariate Time Trajectories.* Proceedings of COMPSTAT 2006, Roma, Italy **1**, 17–29 (2006)
12. V. Faber, Clustering and the continuous k-means algorithm. Los Alamos Sci. **22**, 138–144 (1994)
13. O. Seref, Y.-J. Fan, W.A. Chaowalitwongse, Mathematical programming formulations and algorithms for discrete k-median clustering of time series data. INFORMS J. Comput., *Forthcoming*
14. D. Gada, K.K. Dhiral, K. Kalpakis, V. Puttagunta, in *Distance Measures for Effective Clustering of Arima Time-Series.* Proceedings of the 2001 IEEE International Conference on Data Mining, San Jose, California, pp. 273–280 (2001)
15. M.R. Garey, D.S. Johnson, *Computers and Intractibility: A Guide to the Theory of NP-Completeness* (W. H. Freeman, CA, 1979)
16. K. Jain, V.V. Vazirani, Approximation algorithms for metric facility location and k-median problems using the primal-dual schema and lagrangian relaxation. J. ACM **48**(2), 274–296 (2001)
17. A. Joshi, R. Krishnapuram, L. Yi, in *A Fuzzy Relative of the k-Medoids Algorithm with Application to Web Document and Snippet Clustering.* Snippet Clustering, Proceedings of IEEE International Conference on Fuzzy Systems – FUZZ-IEEE99, Seoul, Korea, (1999)
18. O. Kariv, S.L. Hakimi, An algorithmic approach to network location problems. ii: The p-medians. SIAM J. Appl. Math. **37**(3), 539–560 (1979)
19. L. Kaufman, P.J. Rousseeuw, *Finding Groups in Data: An Introduction to Cluster Analysis (Wiley Series in Probability and Statistics)* (Wiley-Interscience, NY, 2005)
20. E. Keogh, C.A. Ratanamahatana, Exact indexing of dynamic time warping. Knowl. Inform. Syst. **7**(3), 358–386 (2005)
21. S.S. Khan, A. Ahmad, Cluster center initialization algorithm for k-means clustering. Pattern Recogn. Lett. **25**(11), 1293–1302 (2004)
22. J. Liang, H. Zhao, G. Zhang, in *Fuzzy k-Median Clustering Based on hsim Function for the High Dimensional Data.* Proceedings of the 6th World Congress on Intelligent Control and Automation, Dalian, China, pp. 3099–3102 (2006)
23. S.P. Lloyd, Least squares quantization in pcm. IEEE Trans. Inform. Theor. **28**, 129–137 (1982)
24. O.L. Mangasarian P.S. Bradley, W.N. Street, Clustering via concave minimization. Adv. Neural Inform. Process. Syst. **9**, 368–374 (1997)
25. J.-P. Mei, L. Chen, Fuzzy clustering with weighted medoids for relational data. Pattern Recogn. **43**, 1964–1974 (2010)
26. M.N. Murty, A.K. Jain, P.J. Flynn, Data clustering: A review. ACM Comput. Surv. **31**, 264–323 (1999)
27. O. Nasraoui, R. Krishnapuram, A. Joshi, L. Yi, Low-complexity fuzzy relational clustering algorithms for webmining. IEEE Trans. Fuzzy Syst. **9**, 595–607 (2001)
28. K. Pollard, M. Van Der Laan, J. Bryan, A new partitioning around medoids algorithm. J. Stat. Comput. Simulation **73**(8), 575–584 (2003)
29. P. Raghavan, C.D. Manning, H. Schütze, *Introduction to Information Retrieval* (Cambridge University Press, London, 2008)
30. C.S. Revelle, R.W. Swain, Central facilities location. Geogr. Anal. **2**(1), 30–42 (1970)
31. P.P. Rodrigues, J. Gama, J. Pedroso, Hierarchical clustering of time-series data streams. IEEE Trans. Knowl. Data Eng. **20**, 615–627 (2008)
32. P.H.A. Sneath, R.R. Sokal, *Numerical Taxonomy: The Principles and Practice of Numerical Classification* (W.H. Freeman, San Francisco, 1973)
33. E. Tardos, M. Charikara, S. Guhab, D.B. Shmoys, A constant-factor approximation algorithm for the k-median problem. J. Comp. Syst. Sci. **65**(1), 129–149 (2002)

34. N. Vlassis, A. Likas, J.J. Verbeek, The global k-means clustering algorithm. Pattern Recogn. **36**, 451–461 (2001)

35. L. Wei, E. Keogh, X. Xi, C.A. Ratanamahatana, The UCR Time Series Classification/Clustering Homepage, http://www.cs.ucr.edu/~eamonn/time_series_data/(2006)

36. X. Xi, S.H. Lee, E. Keogh, L. Wei, M. Vlachos, in *Lb_keogh Supports Exact Indexing of Shapes Under Rotation Invariance with Arbitrary Representations and Distance Measures.* VLDB '06: Proceedings of the 32nd International Conference on Very Large Data Bases (VLDB Endowment, 2006), Seoul, Korea, pp. 882–893 (2006)

Mathematical Models of Supervised Learning and Application to Medical Diagnosis

Roberta De Asmundis and Mario Rosario Guarracino

Abstract Supervised learning models are applicable in many fields of science and technology, such as economics, engineering and medicine. Among supervised learning algorithms, there are the so-called Support Vector Machines (SVMs), exhibiting accurate solutions and low training time. They are based on the statistical learning theory and provide the solution by minimizing a quadratic type cost function. SVMs, in conjunction with the use of kernel methods, provide non-linear classification models, namely separations that cannot be expressed using inequalities on linear combinations of parameters. There are some issues that may reduce the effectiveness of these methods. For example, in multi-center clinical trials, experts from different institutions collect data on many patients. In this case, techniques currently in use determine the model considering all the available data. Although they are well suited to cases under consideration, they do not provide accurate answers in general. Therefore, it is necessary to identify a subset of the training set which contains all available information, providing a model that still generalizes to new testing data. It is also possible that the training sets vary over time, for example, because data are added and modified as a result of new tests or new knowledge. In this case, the current techniques are not able to capture the changes, but need to start the learning process from the beginning. The techniques, which extract only the new knowledge contained in the data and provide the learning model in an incremental way, have the advantage of taking into account only the

R. De Asmundis (✉)
Department of Statistical Sciences (DSS), University of Rome 'La Sapienza', Rome, Italy
e-mail: roberta.deasmundis@uniroma1.it

M.R. Guarracino
High Performance Computing and Networking Institute, Italian National
Research Council, Naples, Italy
e-mail: mario.guarracino@cnr.it

P.M. Pardalos et al. (eds.), *Optimization and Data Analysis in Biomedical Informatics*,
Fields Institute Communications 63, DOI 10.1007/978-1-4614-4133-5_3,
© Springer Science+Business Media New York 2012

really useful experiments and speed up the analysis. In this paper, we describe some solutions to these problems, with the support of numerical experiments on the discrimination among differ types of leukemia.

Mathematics Subject Classification (2010): Primary 68T10, Secondary 62H30

1   Introduction

The genetic information of an organism is stored in DNA molecules which contain four types of *nucleotides* to compose the genome of each living organism. Parts of the genome, called *genes*, have the capability to transcribe RNA, through a process called *gene expression*. As for DNA, RNA is composed of nucleotides to encode genetic information and all living organism use messenger RNA (mRNA) to carry the genetic information that directs the synthesis of proteins. The products of gene expression are molecules composed of RNA, from which proteins are translated. In translation, triplets of nucleotides (codons) determine which will be the next amino acid added in the growing protein chain. The sequence between the starting and ending encoding nucleotides is called an open reading frame. The human genome contains more than three billions base pairs, but the complexity of a genome is not directly related to the complexity of an organism. Indeed, the latter is connected to the differentiation and specialization of cells and their capability to communicate and interact to perform complex tasks.

In 1995 Schena et al. [1] developed a microarray system to simultaneously measure the level of expression of 45 genes in the flowering plant Arabidopsis. This was a major improvement with respect to existing techniques, that could report the activity of single genes. Tens of thousands publications have appeared in the scientific literature since 1995, highlighting the rapid proliferation of microarray technology. This technology is nowadays widely used in both medical and biological research. The analysis of microarray data can detect the expression of tens of thousands of genes in a single experiment.

DNA microarrays are typically glass slides on which tens of thousands spots of DNA are printed. Each spot corresponds to some portion of a known gene or predicted open reading frame. Each spot identifies, through a process called hybridization, the expression level of the mRNA transcript by a gene. The output of this process is a digital image, that contains a color spot for each probe, in each experiment, as depicted in Fig. 1. To obtain information about expression levels, each spot is identified and its intensity measured. This process is prone to errors. First, it may happen the spots are not perfectly aligned, and therefore, a registration phase is needed to assign spots to probes. Then, the spot color is not uniform, and from its intensity a single number has to be produced. In this process many values are missing or computed with an error. In Affymetrix mycroarrays, for example, the gene expression value is paired with a "P", (present), if the value could be evaluated with enough confidence, a "A" (absent) if the value is absent, or an "M" (marginal) if the the statistical significance of the expression value provided is low.

Fig. 1 Detail of a typical
example of 40.000 probes
microarray

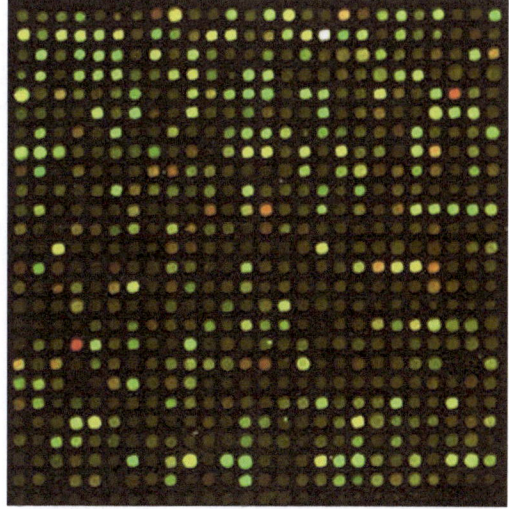

We have now a very large quantity of microarray experiments stored in public databases, such as GEO [2] at the National Center for Biotechnology Information, or ArrayExpress [3] at the European Bioinformatics Institute. These data have been collected and annotated, and they grow at unbelievable speed, requiring a daily rebuild. In the field of medical diagnosis, microarrays have been used for many different purposes: Alon et al. [4] propose to classify tumor versus normal colon tissues, Golub et al. [5] to discriminate between Acute Myeloid versus Lymphoblastic Leukemia, Hedenfalk et al. [6] to classify BRCA1 vs. BRCA2 and sporadic mutations, Singh et al. [7] to predict patient outcome after prostatectomy; Veer et al. [8] to predict the clinical outcome of breast cancer, Nutt et al. [9] to derive a prognosis on gliomas survival, and Iizuka et al. [10] to predict the recurrence of hepatocellaur carcinoma after curative resection.

Microarrays are currently the most popular technique for transcript profiling and their cost can be afforded by many laboratories. Unfortunately, there are some limits, that are directly connected with the acquisition process and technology. First, they depend on prior probe selection and can have a limited sensitivity due to the introduction of biases during the hybridization process [11]. Then, during the statistical analysis, problems arise from the large size of datasets to be processed, the large number of variables (curse of dimensionality), and the updating of training sets.

In this paper we propose a solution to the described problems, based on incremental construction of the training sets and a decremented characterization of the probes involved in the classification of data. This strategy will provide benefits with respect to the computational time needed to compute and update the classification models and accuracy of solutions.

The present work is organized as follow: we describe the *de facto* standard classification algorithms and the incremental strategy in Sect. 2; we introduce the

Gene Selection technique and we present a case study applied to the classification among different types of leukemia in Sect. 3; Sect. 4 contains the description of some future directions such as deep sequencing, system biology and personalized healthcare; fineally we give our conclusions in Sect. 5.

The notation used in the paper is as follows. All vectors are column vectors and are denoted in bold, the transposed of a vector $\mathbf{x}$ will be indicated by $\mathbf{x}'$ and the transposed of a matrix A by A'. Scalar product of two vectors $\mathbf{x}$ and $\mathbf{y}$ in $\mathbf{R}^n$ will be denoted by $\mathbf{x}'\mathbf{y}$, 2-norm of $\mathbf{x}$ will be denoted by $\|\mathbf{x}\|$ and $\mathbf{e}$ is a vector of ones of appropriate size.

2 Supervised Learning

2.1 Support Vector Machines

Support Vector Machines (SVMs) are the state-of-the-art supervised classification methods, widely accepted in many application areas. A SVM finds an hyperplane $\mathbf{w}'\mathbf{x} + b = 0$ with the objective to separate the elements belonging to two different classes [12]. The separating hyperplane is usually chosen to maximize the *margin* between the two classes, which can be defined as the minimum distance of all of the elements of either class to the hyperplane and it is equal to:

$$\mu = \frac{2}{\| \mathbf{w} \|}. \tag{1}$$

The elements which realize the maximum margin are called *support vectors* and are the only elements needed to train the classifier because [13] the weights $\mathbf{w}$ for the optimal hyperplane can be written as some liner combination of support vectors:

$$\mathbf{w} = \sum_i \alpha_i y_i \mathbf{x}_i. \tag{2}$$

Figure 2 shows the hyperplane that separates the points of the two classes, the margin μ and the support vectors which are linked to the optimal hyperplane.

Let consider a data set composed of k pairs $(\mathbf{x}_i, y_i)$ where $\mathbf{x}_i \in \mathbf{R}^n$ is the feature vector that characterizes the point $\mathbf{x}_i$, and $y_i \in \{-1, 1\}$, is the class label. Then, the solution to the following quadratic linearly constrained problem is the optimal hyperplane with the maximum margin:

$$\min f(\mathbf{w}) = \frac{\mathbf{w}'\mathbf{w}}{2} \tag{3}$$

$$s.t. \ \mathbf{w}'\mathbf{x}_i + b \geq 1 \quad y_i \in class1$$

$$\mathbf{w}'\mathbf{x}_i + b \leq -1 \quad y_i \in class2. \tag{4}$$

Fig. 2 A separable problem
in a 2 dimensional space

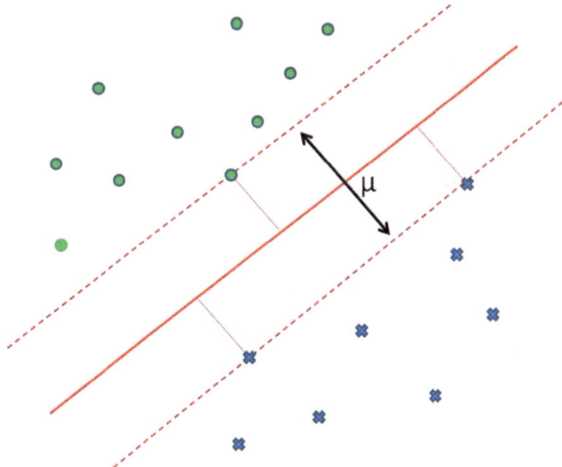

Note that the objective function of (3) is different from (1), this is due to mathematical reasons in order to simplify the problem without changing the solution. The constraints (4) can be simplified to a single expression:

$$y_i(\mathbf{w}'\mathbf{x}_i + b) \geq 1.$$

A dual representation of this problem can be given as follows:

$$\max f(\alpha) = \frac{1}{2}\alpha' Q\alpha - \mathbf{e}'\alpha$$

$$s.t. \quad \alpha_i \geq 0, i = 1,\ldots,n$$

$$\mathbf{y}'\alpha = 0, \tag{5}$$

and the classification function is:

$$f(\mathbf{x}) = \sum_{i=1}^{n} \alpha_i y_i \mathbf{x}_i \mathbf{x}, \tag{6}$$

where α is a vector of Lagrange multipliers and $Q_{i,j} = y_i y_j (\mathbf{x}_i'\mathbf{x}_j)$.

Considering two matrices $A \in \mathbf{R}^{p \times n}$ and $B \in \mathbf{R}^{m \times n}$, that represent the two classes, each row being a point in the input space, the quadratic linearly constrained problem (3), which has to be solved to obtain the optimal hyperplane, identified by $(\mathbf{w}, b)$, can also be written as:

$$\min f(\mathbf{w}) = \frac{\mathbf{w}'\mathbf{w}}{2}$$

$$s.t. \quad (A\mathbf{w} + b) \geq \mathbf{e}$$

$$(B\mathbf{w} + b) \leq -\mathbf{e}. \tag{7}$$

Fig. 3 A non separable
problem in a 2 dimensional
space; here the right way to
separate classes is with a
circle

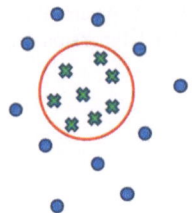

The advantage of SVMs is that the optimal hyperplane can easily be found using a
small subset of points from both classes whose memberships are a priori known and
the test points can be efficiently classified using the function $f(\mathbf{x})$ in (6).

In some cases it is impossible to find a separating hyperplane because the data
can be nonlinearly separable, Fig. 3.

In such a case, the initial sets of points representing the data, which originally
reside in the *input space*, can be nonlinearly embedded to a higher dimensional
space, called *feature space*, in which the linear separation, and so the optimal
hyperplane, can be found. This nonlinear mapping can be implicitly done by kernel
functions [14], which represent the inner product of the elements in the nonlinear
space. In this way the classifier is a hyperplane in the higher-dimensional feature
space, but it can be nonlinear in the original input space.

Some common kernels are the following:

Gaussian Radial Basis Function:

$$K(\mathbf{x}_i, \mathbf{x}_j) = e^{-\frac{\|\mathbf{x}_i - \mathbf{x}_j\|^2}{\sigma}}. \tag{8}$$

Polynomial homogeneous:

$$K(\mathbf{x}_i, \mathbf{x}_j) = (\mathbf{x}_i \mathbf{x}_j)^d. \tag{9}$$

Polynomial inhomogeneous:

$$K(\mathbf{x}_i, \mathbf{x}_j) = (\mathbf{x}_i \mathbf{x}_j + 1)^d. \tag{10}$$

Hyperbolic tangent:

$$K(\mathbf{x}_i, \mathbf{x}_j) = \tanh(\mu \mathbf{x}_i \mathbf{x}_j + c)^d, \tag{11}$$

for some $\mu > 0$ and $c < 0$.

The kernel is related to the transform $\phi(\mathbf{x}_i)$ by the equation

$$K(\mathbf{x}_i, \mathbf{x}_j) = \phi(\mathbf{x}_i) \cdot \phi(\mathbf{x}_j), \tag{12}$$

where $\mathbf{x}_i$ and $\mathbf{x}_j$ denote two points in the original input space.

Fig. 4 An example of ReGEC classification applied to a 2 dimensional problem with linearly separable data. The two hyperplane are each the closest to one class and the furthest to the other

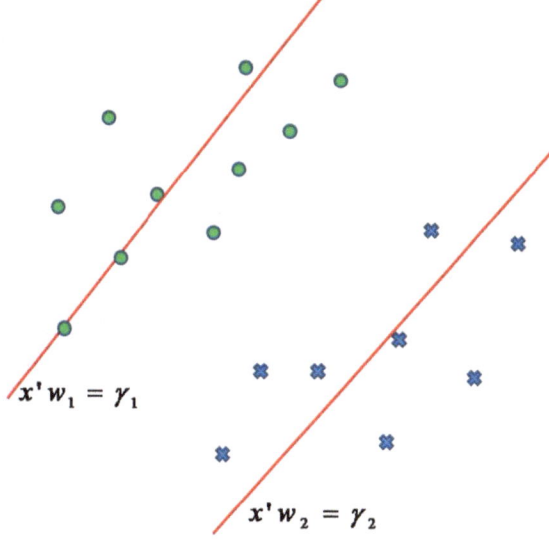

In this way we have:

$$\mathbf{w} = \sum_i \alpha_i y_i \phi(\mathbf{x}_i), \tag{13}$$

and the classification function becomes:

$$f(\mathbf{x}) = \mathbf{w} \cdot \phi(\mathbf{x}) = \sum_i \alpha_i y_i K(\mathbf{x}_i, \mathbf{x}). \tag{14}$$

2.2 ReGEC

Generalized Proximal SVMs (GEPSVM) is another efficient classification algorithm in which the binary classification problem can be formulated as a generalized eigenvalue problem. Mangasarian et al. [15] proposed to classify the two sets of points A and B using two non parallel hyperplanes, each the closest to one set of points and the furthest from the other, as showed in Fig. 4. The assignment of a point to a class is then based on its distance from the two hyperplanes.

For class A, the hyperplane can be obtained by solving the following optimization problem:

$$\min_{\mathbf{w}, \gamma \neq 0} \frac{\|A\mathbf{w} - \mathbf{e}\gamma\|^2}{\|B\mathbf{w} - \mathbf{e}\gamma\|^2}, \tag{15}$$

while the hyperplane for cases in B can be obtained by minimizing the inverse of the objective function in (15). With the following positions:

$$G = [A \quad -\mathbf{e}]'[A \quad -\mathbf{e}],$$
$$H = [B \quad -\mathbf{e}]'[B \quad -\mathbf{e}],$$
$$\mathbf{z} = [\mathbf{w}' \quad \gamma]', \tag{16}$$

where $[A \quad -\mathbf{e}]$ is the matrix obtained from A adding the column vector $-e$, and so for $[B \quad -\mathbf{e}]$; the problem (15) becomes:

$$\min_{\mathbf{z} \neq 0} \frac{\mathbf{z}'G\mathbf{z}}{\mathbf{z}'H\mathbf{z}}. \tag{17}$$

This expression is the Raleigh quotient of the generalized eigenvalue problem:

$$G\mathbf{z} = \lambda H\mathbf{z}. \tag{18}$$

The inverse of the objective function in (17) has the same eigenvectors and reciprocal eigenvalues. Let:

$$\mathbf{z}_{min} = [\mathbf{w}'_1 \quad \gamma_1]', \quad \mathbf{z}_{max} = [\mathbf{w}'_2 \quad \gamma_2]'$$

be the eigenvectors related to the eigenvalues of smallest and largest modulo, respectively. Then, $\mathbf{x}'\mathbf{w}_1 - \gamma_1 = 0$ is the closest hyperplane to the set of points in A and the furthest from those in B, in the same way, $\mathbf{x}'\mathbf{w}_2 - \gamma_2 = 0$ is the closest hyperplane to the set of points in B and the furthest from those in A.

GePSVM can also classify problems that are not linearly separable using kernel functions. The kernel function $K(\mathbf{x}, C) : \mathbf{R}^n \times \mathbf{R}^{(p+m) \times n} \to \mathbf{R}^{p+m}$ can embed the points in a higher dimensional space, thus the resulting hyperplane, projected in the feature space [16], has equation:

$$K(\mathbf{x}, C)\mathbf{u} - \gamma = 0. \tag{19}$$

In this study we use the *Gaussian kernel* (8), where $\mathbf{x}_i$ and $\mathbf{x}_j$ denote two points in the original input space.

The nonlinear implicit mapping is done through the *kernel matrix* $K(A, B)$, whose elements are defined as:

$$K(A, B)_{i,j} = e^{-\frac{\|A_i - B_j\|^2}{\sigma}}, \tag{20}$$

where A_i and B_j are the ith and jth rows of the matrices A and B, respectively. We now look for two hyperplanes of equations:

$$K(\mathbf{x}, C)\mathbf{u}_1 - \gamma_1 = 0, \quad K(\mathbf{x}, C)\mathbf{u}_2 - \gamma_2 = 0, \tag{21}$$

each closest to one set of points, and furthest from the other.

The formulation of the optimization problem required to be solved using kernel function is the following:

$$\min_{u,\gamma\neq 0} \frac{\|K(A,C)\mathbf{u} - \mathbf{e}\gamma\|^2}{\|K(B,C)\mathbf{u} - \mathbf{e}\gamma\|^2}. \tag{22}$$

Substituting the scalar product expression of the norms in the above equation yields:

$$\min_{u,\gamma\neq 0} \frac{[K(A,C)\mathbf{u} - \mathbf{e}\gamma]'[K(A,C)\mathbf{u} - \mathbf{e}\gamma]}{[K(B,C)\mathbf{u} - \mathbf{e}\gamma]'[K(B,C)\mathbf{u} - \mathbf{e}\gamma]}. \tag{23}$$

With the following positions:

$$G = [K(A,C) \ -\mathbf{e}\gamma]'[K(A,C) \ -\mathbf{e}\gamma]$$
$$H = [K(B,C) \ -\mathbf{e}\gamma]'[K(B,C) \ -\mathbf{e}\gamma],$$

with $\mathbf{z}' = [\mathbf{u}' \ \gamma]$, (23) can be rewritten as:

$$\min_{z\neq 0} \frac{\mathbf{z}'G\mathbf{z}}{\mathbf{z}'H\mathbf{z}}. \tag{24}$$

As above, this is the Rayleigh quotient of the generalized eigenvalue problem $G\mathbf{z} = H\lambda\mathbf{z}$. When H is positive definite, the stationary points of (24) are achieved at the eigenvectors in which the objective function is equal to the corresponding eigenvalue. This means that the solution to (24) is achieved at the eigenvector with the minimum eigenvalue.

Since the matrices G and H can be deeply rank deficient, there is the possibility that the null spaces of the two matrices have a non trivial intersection. This is due to the fact that when G or H are not full rank, problems (17) and (24) can have multiple eigenvectors related to a single eigenvalue. For example, when H has a zero eigenvalue of multiplicity r, the generalized eigenvalue problem has r eigenvectors related to the 'infinite' eigenvalues. This leads to a problem that can be ill-conditioned and therefore a regularization technique needs to be applied in order to numerically solve the problem.

Mangasarian proposed to solve the following two regularized optimization problems, where $C' = [A' \ B']$ and δ is a regularization parameter:

$$\min_{u,\gamma\neq 0} \frac{\|K(A,C)\mathbf{u} - \mathbf{e}\gamma\|^2 + \delta\|\begin{bmatrix} u \\ \gamma \end{bmatrix}\|^2}{\|K(B,C)\mathbf{u} - \mathbf{e}\gamma\|^2} \tag{25}$$

and

$$\min_{u,\gamma\neq 0} \frac{\|K(B,C)\mathbf{u} - \mathbf{e}\gamma\|^2 + \delta\|\begin{bmatrix} u \\ \gamma \end{bmatrix}\|^2}{\|K(A,C)\mathbf{u} - \mathbf{e}\gamma\|^2}. \tag{26}$$

The number of eigenvalue problems can be reduced from two to one, using the new regularization method *ReGEC*, proposed by Guarracino et al. [17], by solving the following generalized eigenvalue problem:

$$\min_{\mathbf{u},\gamma \neq 0} \frac{\|K(A,C)\mathbf{u} - \mathbf{e}\gamma\|^2 + \delta\|\tilde{K}_B\mathbf{u} - \mathbf{e}\gamma\|^2}{\|K(B,C)\mathbf{u} - \mathbf{e}\gamma\|^2 + \delta\|\tilde{K}_A\mathbf{u} - \mathbf{e}\gamma\|^2}. \tag{27}$$

Here $\tilde{K}_A$ and $\tilde{K}_B$ are diagonal matrices whose entries are the main diagonals of the kernel matrices $K(A,C)$ and $K(B,C)$ respectively. The new regularization provides classification accuracy results comparable to the ones obtained by solving equations (25) and (26) and it is a form of *robustification*.

2.3 Incremental

Incremental classification has been introduced to reduce the training data to a substantially small and robust subset, providing comparable accuracy results. The use of a smaller set of points reduces the probability of over-fitting the problem and is computationally easier to handle than the original set.

Incremental Classification with Generalized Eigenvalues [18], I-REGEC, also provides a constructive way to understand the influence of new training data on an existing classification function. As new points become available, the cost of retraining the algorithm decreases if the influence of the new points is only evaluated with respect to the small subset, rather than the whole training set.

The algorithm takes an initial set of points C_0 and the entire training set C as input, such that $C \supset C_0 = A_0 \cup B_0$, and A_0 and B_0 are sets of points in C_0 that belong to the two classes A and B. We refer to C_0 as the *incremental subset*. Let $\Gamma_0 = C \setminus C_0$ be the initial set of points that can be included in the incremental subset. ReGEC classifies all of the points in the training set C using the kernel from C_0. Let P_{A_0} and P_{B_0} be the hyperplanes found by ReGEC, Acc_0 be the classification accuracy and M_0 be the points that are misclassified. Then, among the points in $\Gamma_0 \cap M_0$ the point that is farthest from its respective hyperplane is selected, i.e.,

$$\mathbf{x}_1 = \mathbf{x}_i : \max_{\mathbf{x} \in \{\Gamma_0 \cap M_0\}} \{dist(\mathbf{x}, P_{class(\mathbf{x})})\}, \tag{28}$$

where $class(\mathbf{x})$ returns A or B depending on the class of x. This point is the candidate point to be included in the incremental subset. This choice is based on the idea that a point very far from its plane may be needed in the classification subset in order to improve accuracy. We update the incremental set as $C_1 = C_0 \cup \{\mathbf{x}_1\}$. Then, we classify the entire training set C using the points in C_1 to build the kernel.

Algorithm 1 I-ReGEC(C_0, C)

1: $\Gamma_0 = C \setminus C_0$
2: $\{M_0, Acc_0\} = Classify(C, C_0)$
3: $k = 1$
4: **while** $|\Gamma_k| > 0$ **do**
5:     $\mathbf{x}_k = \mathbf{x} : \max_{\mathbf{x} \in \{M_{k-1} \cap \Gamma_{k-1}\}} \{dist(\mathbf{x}, P_{class(\mathbf{x})})\}$
6:     $\{M_k, Acc_k\} = Classify(C, \{C_{k-1} \cup \{\mathbf{x}_k\}\})$
7:     **if** $Acc_k > Acc_{k-1}$ **then**
8:        $C_k = C_{k-1} \cup \{\mathbf{x}_k\}$
9:     **end if**
10:    $\Gamma_k = \Gamma_{k-1} \setminus \{\mathbf{x}_k\}$
11:    $k = k + 1$
12: **end while**

Let the classification accuracy be Acc_1. If $Acc_1 > Acc_0$ then we keep the new subset; otherwise we reject the new point, that is $C_1 = C_0$. In both cases $\Gamma_1 = \Gamma_0 \setminus \{x_1\}$. The algorithm repeats until the condition $|\Gamma_k| = 0$ is reached. Algorithm 1 is a pseudo-code of the described method.

3 Gene Expression Data Analysis

Gene expression is the process by which information from a gene is used in the synthesis of a functional gene product. These products are often proteins or a functional RNA. Gene expression analysis is applied in medicine to scan the expression levels of tens of thousands of genes using a microarray, in order to classify patients in different groups. In this way is possible to classify types of cancers with respect to the patterns of gene activity in the tumor cells. The information needed to classify samples is often contained in patterns involving few tens of genes, but selecting those useful biomarkers is a nontrivial task.

3.1 Gene Selection

Techniques for making easier the biomarkers selection can be divided in two groups: standard and statistical methods. Standard methods, such as RFE or Relief, need long and memory intensive computations, while statistical techniques, such as LDA or FDA, are much faster, but can produce low accuracy results [19], that is why there is the need for hybrid techniques that can take advantage of both approaches. Our technique consists in a simultaneous incremental learning and decremented characterization (ILDC), which permit to acquire knowledge on gene grouping during the classification process. This technique relies on standard statistical indexes (mean μ and standard deviation σ):

$$F\left(\mathbf{x}_j\right) = \left|\frac{\mu_j^+ - \mu_j^-}{\sigma_j^+ + \sigma_j^-}\right|.$$

We use the values of $F(\mathbf{x}_j)$ to choose among genes, preferring the ones with a greater value of F.

3.2   A Case Study

We consider the Golub microarray dataset [5] composed by 72 samples with 7,129 gene expression values. A principal component analysis divided the dataset in two clusters, the first containing 25 Acute Myeloid Leukemia samples, and the second 47 Acute Lymphoblastic Leukemia samples. Applying ILDC-ReGEC to the dataset we discovered that only 100 genes out of 7,129 are responsible of this discrimination, those 100 selected genes are in agreement with previous studies and less then 10 patients, out of 72, are needed for the training phase. We reached a classification accuracy of 96.86%.

4   Future Directions

The human genome was sequenced 10 years ago. Many steps have been made in technology and in knowledge, but still there are millions of open questions. One of the most exciting advances has been the development of low cost, high-throughput methodologies for studying human genome-scale variations [20]. These technologies can led to the identification of genetic variants with roles in human phenotypic variation, both in relation to disease susceptibility and evolutionary adaptation [21]. The development of statistical approaches for analysing genome-scale variation has also been crucially important.

4.1   Deep Sequencing

Sequencing technologies have also made a strong contribution to understanding human genetic variation. The first two draft human genome sequences were completed in 2001, and the first human genome was resequenced using nextgeneration sequencing (NGS) technology, at a fraction of the cost, in 2008. New studies have underlined that individual genomes may differ by megabases of sequence because of structural variation, including insertions, deletions and inversions. Deep sequencing is a new technology which can produce massive sequencing of billions of DNA sequences in the form of short fragments. This new technology allows to obtain detailed information on any possible genetic alteration found in tumor cells. The application of this technology in cancer discrimination, could lead to the identification of genetic alterations still unknown, opening new horizons in the search of rare disease.

4.2   System Biology

Using high-tech procedures, new disciplines of the post-genomics period have the ambitious goal of achieving an overall vision of all phenomena occurring in the particular circumstances in which an organic system is set. To the set of those disciplines is given the name of *Systems Biology*. The integration of systems biology approaches and studies of natural selection may be particularly informative. Living organisms can be represented as complex networks of interacting molecules, linking such networks will eventually enable a systems-level understanding of living organisms. The prediction of such interactions is important for medical diagnosis and to devise new drugs and some new models capable to describe complex networks are currently in demand.

4.3   Personalized Healthcare

The gene number paradox, according to which organisms of different complexity have similar numbers of protein-coding genes, illustrates how a certain perception of biology revolves around the notion that to understand an organism, we need only to understand all its genes [22]. According to this, if we understand all the genes of an organism, we understand the whole organism itself. Is it also a necessary condition? Sequencing individual genomes is becoming an affordable task in terms of time and cost. Available software technologies store data from a population point of view, making it difficult to manage and analyze complete information about individuals. We would like to move from a global population to a local individual point of view. This paradigm shift will change the way in which we develop data analysis techniques.

5   Conclusion

Technology in biology and biomedicine applications is progressing at an unprecedented speed and scientists need new methodologies to have more insight in their data. These methodologies need to provide models based on large volumes of data, often noisy and incomplete. Results obtained using microarray consist of thousands of data which validity and significance should be evaluated. The need of validating and managing a large amount of informations made it necessary to define a rigorous process of statistical analysis of results and dynamic analysis of experimental data.

Acknowledgments This work has been partially funded by the FLAGSHIP "InterOmics" project (PB.P05), by the Italian MIUR and CNR organizations.

References

1. M. Schena, D. Shalon, R.W. Davis, P.O. Brown, Quantitative Monitoring of Gene Expression Patterns with a Complementary DNA Microarray. **270**(5235), 467–470 (1995)
2. T. Barrett, D.B. Troup, S.E., Wilhite, P. Ledoux, C. Evangelista, I.F. Kim, M. Tomashevsky, K.A. Marshall, K.H. Phillippy, P.M. Sherman, R.N. Muertter, M. Holko, O. Ayanbule, A. Yefanov, A. Soboleva, NCBI GEO: Archive for functional genomics data sets–10 years on. Nucl. Acids Res. **39**, D1005–D1010 (2011)
3. H. Parkinson, U. Sarkans, N. Kolesnikov, N. Abeygunawardena, T. Burdett, M. Dylag, I. Emam, A. Farne, E. Hastings, E. Holloway, N. Kurbatova, M. Lukk, I. Malone, R. Mani, E. Pilicheva, G. Rustici, A. Sharma, E. Williams, T. Adamusiak, M. Brandizi, N. Sklyar, A. Brazma, ArrayExpress update–an archive of microarray and high-throughput sequencing-based functional genomics experiments. Nucleic Acids Res. **39**, D1002-4. Epub. (2011)
4. A. Alon, N. Barkai, D.A. Notterman, K. Gish, S. Ybarra, D. Mack, A.J. Levine, Broad patterns of gene expression revealed by clustering analysis of tumor and normal colon tissues probed by oligonucleotide arrays. Proc. Natl. Acad. Sci. U.S.A. **96**(12), 6745–6750 (1999)
5. Golub et al., Molecular classifcation of cancer: Class discovery and class prediction by gene expression monitoring. Science **286**, 531–537 (1999)
6. I. Hedenfalk, D. Duggan, Y. Chen, M. Radmacher, R. Simon, P. Meltzer, B. Gusterson, M. Esteller, M. Raffeld, Z. Yakhini, A. Ben-Dor, E. Dougherty, J. Kononen, L. Bubendorf, W. Fehrle, S. Pttalunga, S. Gruvberger, N. Loman, O. Johannsson, H. Olsson, B. Wilfond, G. Sauter, O.P. Kallioniemi, A. Borg, J. Trent, Gene-expression profiles in hereditary breast cancer. New Engl. J. Med. **344**, 539–548 (2001)
7. D. Singh, P.G. Febbo, K. Ross, D.G. Jackson, J. Manola, C. Ladd, P. Tamayo, A.A. Renshaw, A.V. D'Amico, J.P. Richie, E.S. Lander, M. Loda, P.W. Kantoff, T.R. Golub, W.R. Sellers, Gene expression correlates of clinical prostate cancer behavior. Cancer Cell **1**(2), 203–209 (2002)
8. L. J. van't Veer, H. Dai, M. van de Vijver, Y. He, A. Hart, M. Mao, H. Peterse, K. van der Kooy, M. Marton, A. Witteveen, G. Schreiber, R. Kerkhoven, C. Roberts, P. Linsley, R. Bernards, S. Friend, Gene expression profiling predicts clinical outcome of breast cancer Nature **415**, 530–536 (2002)
9. C.L. Nutt, D.R. Mani, R.A. Betensky, P. Tamayo, J.G. Cairncross, C. Ladd, U. Pohl, C. Hartmann, M.F. McLaughlin, T.T. Batchelor, P.M. Black, A. von Deimling, S.L. Pomeroy, T.R. Golub, D.N. Louis, Gene expression-based classification of malignant gliomas correlates better with survival than histological classification. Cancer Res. **63**(7), 1602–1607 (2003)
10. N. Iizuka, M. Oka, H. Yamada Okabe, M. Nishida, Y. Maeda, N. Mori, T. Takao, T. Tamesa, A. Tangoku, H. Tabuchi, K. Hamada, H. Nakayama, H. Ishitsuka, T. Miyamoto, A. Hirabayashi, S. Uchimura, Y. Hamamoto, Oligonucleotide microarray for prediction of early intrahepatic recurrence of hepatocellular carcinoma after curative resection. The Lancet **361**, 923–929 (2003)
11. S. Baginsky, L. Henning, P. Zimmermann, W. Gruissem, Gene expression analysis, proteomics, and network discovery. Plant Physiol. **152**, 402–410 (2010); American Society of Plant Biologists

12. V. Vapnik, *The Nature of Statistical Learning Theory* (Springer, New York, 1995)
13. C. Cortes, V. Vapnik, Support-vector networks. Mach. Learn. **20**, 273–297 (1995)
14. B.E. Boser, I.M. Guyon, V.N. Vapnik, *A Training Algorithm for Optimal Margin Classifiers.* 5th Annual ACM Workshop on COLT, Pittsburgh, PA, 1992, pp. 144–152
15. O.L. Mangasarian, E.W. Wild, Multisurface proximal support vector machine classification via generalized eigenvalues. IEEE Trans Pattern Anal Mach Intell.; **28**, 69–74 (2006)
16. B. Schölop, A.J. Smola, *Learning with Kernels: Support Vector Machines, Regularization, Optimization, and Beyond* (MIT, MA, 2001)
17. M.R. Guarracino, C. Cifarelli, O. Seref, P.M. Pardalos, A classification method based on generalized eigenvalue problems. Optim. Meth. Software **22**, 73–81 (2007)
18. C. Cifarelli, M.R. Guarracino, O. Seref, S. Cuciniello, P.M. Pardalos, Incremental classifcation with generalized eigenvalues. J. Class. **24**(2), 205–219 (2007)
19. I. Guyon, A. Elisseeff, An introduction to variable and feature selection. J. Mach. Learn. Res. **3**, 1157–1182 (2003)
20. E.S. Lander et al., Initial sequencing and analysis of the human genome. Nature **409**, 860–921 (2001)
21. D. Wheeler et al., The complete genome of an individual by massively parallel DNA sequencing. Nature **452**, 872–876 (2008)
22. E. Heard, S. Tishkoff, J. Todd, M. Vidal, G. Wagner, J. Wang, D. Weigel, R. Young, Ten years of genetics and genomics: what have we achieved and where are we heading? Nature Reviews Genetics **11**, 723–733 (2010)

Predictive Model for Early Detection of Mild Cognitive Impairment and Alzheimer's Disease

Eva K. Lee, Tsung-Lin Wu, Felicia Goldstein, and Allan Levey

Abstract The number of people affected by Alzheimer's disease is growing at a rapid rate, and the consequent increase in costs will have significant impacts on the world's economies and health care systems. Therefore, there is an urgent need to identify mechanisms that can provide early detection of the disease to allow for timely intervention. Neuropsychological tests are inexpensive, non-invasive, and can be incorporated within an annual physical examination. Thus they can serve as a baseline for early cognitive impairment or Alzheimer's disease risk prediction. In this paper, we describe a PSO-DAMIP machine-learning framework for early detection of mild cognitive impairment and Alzheimer's disease. Using two trials of patients with Alzheimer's disease (AD), mild cognitive impairment (MCI), and control groups, we show that one can successfully develop a classification rule based on data from neuropsychological tests to predict AD, MCI, and normal subjects where the blind prediction accuracy is over 90%. Further, our study strongly suggests that raw data of neuropsychological tests have higher potential to predict subjects from AD, MCI, and control groups than pre-processed subtotal score-like features. The classification approach and the results discussed herein offer the potential for development of a clinical decision making tool. Further study must be conducted to validate its clinical significance and its predictive accuracy among various demographic groups and across multiple sites.

E.K. Lee (✉) • T.-L. Wu
Center for Operations Research in Medicine and HealthCare, NSF I/UCRC Center for Health Organization Transformation, and School of Industrial and Systems Engineering, Georgia Institute of Technology, Atlanta, GA 30332-0205, USA
e-mail: eva.lee@gatech.edu; tlwu@isye.gatech.edu

F. Goldstein • A. Levey
Center for Neurodegenerative Disease and Alzheimer's Disease Center, and Department of Neurology, Emory University, Atlanta, GA 30322, USA
e-mail: felicia.goldstein@emory.edu; allan.levey@emory.edu

P.M. Pardalos et al. (eds.), *Optimization and Data Analysis in Biomedical Informatics*, Fields Institute Communications 63, DOI 10.1007/978-1-4614-4133-5_4, © Springer Science+Business Media New York 2012

83

Mathematics Subject Classification (2010): Primary 90-08, 90C06, 90C11, 90C90, Secondary 92C99

1 Introduction

Alzheimer's disease (AD), the 6th leading cause of death in the United States, is a progressive and irreversible brain disease which causes memory loss and other cognitive problems severe enough to affect daily life. It is estimated that 1 in 8 older Americans and as many as 5.4 million people live with the memory-destroying illness, translating to someone developing AD every 69 s (Alzheimer's Association, 2011). The number of people with Alzheimer's disease is briskly rising, with an estimated 35 million people worldwide currently living with Alzheimer's and other forms of dementia. Currently, AD is incurable; drugs are used to manage the symptoms or to prevent or slow the progress of the disease. Dementia triples health care costs for those over the age of 65, costing over 183 billion dollars annually. The increasing personal and societal costs will have significant impact on the world's economies and health care systems.

Mild cognitive impairment (MCI) is a condition in which there is clear evidence of cognitive problems, most often involving short term memory, but normal day to day functioning is preserved. In other words, MCI is a situation between normal aging and dementia. People with MCI may or may not develop dementia in the future, but people with MCI are at higher risk of developing dementia than those without MCI.

Fifteen years ago, Alzheimer's disease was only accurately diagnosed after death, when doctors performed an autopsy to examine changes in brain tissue. Now, with advances in imaging, biomarkers can help doctors to identify risks of Alzheimer's disease earlier. The evaluation of AD or MCI depends on some clinical and patient data, including complete medical history, neurological exam, laboratory tests, blood tests, neuropsychological tests, brain scans (CT or MRI), and information from close family members. Changes in the brain triggered by Alzheimer's disease develop slowly over many years, thus, the race is on to identify new and non-invasive ways to help diagnose Alzheimer's disease early, even before any symptoms occur.

Non-invasive tests that can identify people who are at-risk but currently have no symptoms are critical to curtail the rapid rise of this illness. Such patients can then be treated so that symptoms never emerge or the onset of symptoms is delayed. Much research has been done in early detection of AD. Several recent studies on detection of early Alzheimer's disease focus on using imaging data [6, 18, 21, 22]. And some new studies also focus on using biomarkers [5, 25, 26]. In this work, we explore the use of neuropsychological data as an early disease prediction marker. Neuropsychological tests are simple, non-invasive, and can be added as a regular routine test during an annual physical examination. If the results can be used to predict the risk-factors of developing Alzheimer's disease, it has the potential to allow for effective detection and early intervention.

Neuropsychological changes in the expression of cognitive declines are important to the diagnosis of AD and MCI. Bondi et al. reviewed neuropsychological changes during the prodromal period of Alzheimer's disease, which are important to the early identification of the disease [2]. Nelson and O'Connor reviewed mild cognitive impairment from the neuropsychological perspective, including the MCI diagnostic criteria, MCI subtypes, and neuropsychological tests, for the purpose of early identification of Alzheimer's disease [24]. The neuropsychological tests which follow certain criteria are good instruments for evaluating neuropsychological status.

Statistical analyses have been applied to neuropsychological data to understand MCI patients. Lopez et al. analyzed neuropsychological characteristics of normal subjects, MCI-amnestic type (MCI-AT) subjects, and MCI-multiple cognitive deficits type (MCI-MCDT) subjects [19]. Tabert et al. conducted hypothesis testing to compare (1) MCI patients with controls, and (2) MCI patients who converted to AD with MCI patients who did not, in a follow-up duration [30].

Besides statistical analyses, some classification models are also applied to neuropsychological data for predictive analysis. Stuss and Trites applied discriminant function analysis to discriminate the control group, the brain-damaged group with a positive physical neurological exam, and the brain-damaged group with a negative result of the same exam [29]. Kluger et al. applied logistic regression and stepwise entry procedure to predict (1) whether nondemented elderly subsequently declined to any diagnosis of dementia; and (2) whether nondemented elderly subsequently declined to a diagnosis of probable Alzheimer's disease [11]. Possible predictor variables included demographic variables, Global Deterioration Scale (GDS) score, and nine cognitive test scores from the neuropsychological battery of NYU Aging and Dementia Research Center.

Our work focuses on using the neuropsychological data to understand the cognitive status of the individuals. In particular, we focus on identifying discriminatory attributes among these data that will allow one to predict individuals with normal brain functioning, mild cognitive impairment, and those with Alzheimer's disease. To demonstrate its accuracy and potential as a clinical decision tool, we use a set of subjects for establishing the predictive rule and perform blind tests on a set of independent subjects to gauge its predictive power.

2 Neuropsychological Data

2.1 Raw Data from Emory Clinical Trials

Anonymous data of neuropsychological tests from 35 subjects were collected at Emory Alzheimer's Disease Research Center from 2004 to 2007. Eighteen types of neuropsychological tests were applied to the subjects, but only four of them were applied to all subjects, thus being used in our predictive model. These tests included

Table 1 Number of subjects
of three groups from two
trials of patients

	AD	MCI	Ctl	Total
Trial 1	5	3	2	10
Trial 2	2	13	10	25
Total	7	16	12	35

1. Mini Mental State Examination (MMSE),
2. Clock drawing test,
3. Word list memory tasks by the Consortium to Establish a Registry for Alzheimer's Disease (CERAD),
4. Geriatric depression scale (GDS).

The MMSE is a screening tool for cognitive impairment, which is brief, but covers five areas of cognitive function, including orientation, registration, attention and calculation, recall, and language. The clock drawing test assesses cognitive functions, particularly visuo-spatial abilities and executive control functions. The CERAD word list memory tasks assess learning ability for new verbal information. The tasks include word list memory with repetition, word list recall, and word list recognition. The GDS is a screening tool to assess the depression in older population.

There were 153 features, including raw data from the four neuropsychological tests as well as subjects' age. Raw data from tests contained answers to individual questions in the tests. Discarding features which contained missing values or which were undiscriminating (i.e., features which contained almost the same value among all subjects), 100 features were used for feature selection and classification. The clinicians also summarize performance of subtotal scores in different tests, resulting in 9 scores for each patient.

The patient data came from two trials. The number of subjects in the trials is listed in Table 1, in which 'Ctl' represents the control group.

2.2   Data from LONI/ADNI

The Alzheimer's Disease Neuroimaging Initiative (ADNI) data website at Laboratory of Neuro Imaging (LONI), UCLA, includes a repository of clinical and imaging data. Clinical data of several neuropsychological tests are used for classification in this study. The neuropsychological tests include clock drawing test, category fluency test, Boston naming test, and so on. The category fluency test requires the systematic retrieval of hierarchically organized information from semantic memory; the Boston naming test measures the ability to name objects of line drawings.

The data set contained results of neuropsychological tests taken by subjects at several time points; we used the data taken at the baseline time point, i.e., the first time a subject took the tests. Data included 819 subjects and 59 features. Unlike the Emory data, the features were pre-processed score-type ones rather than raw data of the tests. After we handled missing values, 786 subjects and 54 features were left for feature selection and classification. The numbers of AD, MCI, and the control group are 223, 388, and 175, respectively.

3 Predictive Model and Machine Learning Framework

3.1 Optimization-Based Predictive Model

We performed classification via the DAMIP approach, (discriminant analysis via mixed integer programming), first developed in 1997 by Gallagher, Lee, and Patterson [9], which realizes the optimal parameters of Andersons classification model [1]. Besides the ability to handle multi-group classification problems, this model incorporates a reserved judgment region and constraints to limit the misclassification rate. Lee and co-authors have since advanced the theoretical and computational properties of this multi-group classification model and successfully applied it across a broad spectrum of biological and medical applications including the differential diagnosis of the type of squamous diseases; predicting presence/absence of heart disease [12, 14]; genomic analysis and prediction of aberrant CpG island methylation in human cancer [7, 20]; discriminant analysis of motility and morphology data in human lung carcinoma; prediction of ultrasonic cell disruption for drug delivery [17]; identification of tumor shape and volume in treatment of sarcoma [15]; multistage discriminant analysis of biomarkers for prediction of early atherosclerois; fingerprinting of native and angiogenic microvascular networks for early diagnosis of diabetes, aging, macular degeneracy and tumor metastasis; prediction of protein localization sites [12]; and vaccine immunogenicity prediction [13, 23, 28]. In each case, the predictive model yields correct classification rates ranging from 80 to 100%. Further, in all these real applications, beyond reporting the tenfold cross-validation results, the resulting classification rule was also blind tested against new independent data of unknown group identity and resulted in remarkable rates of correct prediction.

Theoretically, DAMIP has some appealing characteristics: (1) the misclassification rates using the DAMIP method are consistently lower than other classification approaches in both simulated data and real-world data; (2) the classification rules from DAMIP appear to be insensitive to the specification of prior probabilities, yet capable of reducing misclassification rates when the number of training observations from each group is different; (3) the DAMIP model generates stable classification rules regardless of the proportions of training observations from each group; and (4) the resulting classification rule is universally consistent, given that the Bayes optimal rule for classification is known [3, 4].

First we introduce the notations used in our methods. Suppose in the data we have n observations from K groups with m features. Let $\mathcal{G} = \{1, 2, \ldots, K\}$ be the set of indices of the groups, $\mathcal{O} = \{1, 2, \ldots, n\}$ be the set of indices of the observations, and $\mathscr{F} = \{1, 2, \ldots, m\}$ be the set of indices of the features. Also, let $\mathscr{O}_k$, $k \in \mathscr{G}$ and $\mathscr{O}_k \subseteq \mathscr{O}$, be the set of indices of observations which belong to group k. Moreover, let $\mathscr{F}_j$, $j \in \mathscr{F}$, be the domain of the jth feature, which could be the space of real, integer, or binary values. The ith observation, $i \in \mathscr{O}$, is represented as $(y_i, \mathbf{x}_i) = (y_i, x_{i1}, \ldots, x_{im}) \in \mathscr{G} \times \mathscr{F}_1 \times \cdots \times \mathscr{F}_m$, where y_i is the group of observation i and $(x_{i1}, \ldots, x_{im})$ is the feature vector of observation i. The classification model finds a function $f : (\mathscr{F}_1 \times \cdots \times \mathscr{F}_m) \mapsto \mathscr{G}$ to predict group from the features.

Anderson's classification model maximizes the probability of correct classification subject to some limits of misclassification probability. Let π_k be the prior probability of group k and $f_k(x)$ be the value of the conditional probability density function for the data point $x \in \mathbb{R}^m$ of group k, $k \in \mathscr{G}$. Also let $\alpha_{hk} \in (0, 1)$, $h, k \in \mathscr{G}$, $h \neq k$, be the predetermined limits of the misclassification probability that data of group h are misclassified to group k. The proposed model seeks a partition $\{R_0, R_1, \ldots, R_K\}$ of $\mathbb{R}^m$, where R_k, $k \in \mathscr{G}$, is the region classified to group k and R_0 is the reserved judgment region, in which the group-assignment decision of data points is reserved. Anderson's model is given as:

$$\max \sum_{k \in \mathscr{G}} \pi_k \int_{R_k} f_k(\mathbf{x}) d\mathbf{x}$$

$$\text{s.t.} \int_{R_k} f_h(\mathbf{x}) d\mathbf{x} \le \alpha_{hk}, \quad \forall h, k \in \mathscr{G}, h \neq k \tag{1}$$

Anderson showed that there exists nonnegative constants λ_{hk}, $h, k \in \mathscr{G}, h \neq k$, such that the optimal decision rule of model (1) is given by

$$R_k = \left\{ \mathbf{x} \in \mathbb{R}^m : L_k(\mathbf{x}) = \max_{h \in \{0\} \cup \mathscr{G}} L_h(\mathbf{x}) \right\}, \ k \in \{0\} \cup \mathscr{G}, \tag{2}$$

where

$$L_0(\mathbf{x}) = 0$$

$$L_k(\mathbf{x}) = \pi_k f_k(\mathbf{x}) - \sum_{h \in \mathscr{G}, h \neq k} \lambda_{hk} f_h(\mathbf{x}), k \in \mathscr{G} \tag{3}$$

This rule is called Anderson's rule. However, the optimal λ's are difficult to find.

Gallagher et al. first proposed mixed integer programming formulations, named DAMIP, for obtaining the optimal values of λ's in Anderson's rule [8, 9] and subsequently introduced efficient heuristics to obtain good feasible solutions [16]. Nonlinear and linear versions of DAMIP from Gallagher et al. are presented below [9]. The binary variable u_{ki} indicates whether observation i is classified to group k or not. The objective function (4a) maximizes the total number of correctly-classified observations. Constraints (4b) define $L_k(\mathbf{x})$ of (3) in Anderson's rule, constraints (4c) and (4d) guarantee the correct value of u_{ki} based on (2), and constraints (4e) model the misclassification limits. The linear version of DAMIP uses constraints (5a)–(5d) to model constraints (4c) of the nonlinear version, in which the variable t_i achieves the value of $\max\{0, L_{ki} : k \in \mathscr{G}\}$. This (linear) version of DAMIP is almost equivalent to nonlinear DAMIP except that DAMIP introduces a small value ε in its formulation to increase the stability of the classification rule derived by DAMIP, as seen in constraints (5b) and (5d).

Nonlinear DAMIP

$$\max \sum_{i \in \mathscr{O}} u_{y_i i} \qquad (4a)$$

$$\text{s.t. } L_{ki} = \pi_k f_k(\mathbf{x}_i) - \sum_{h \in \mathscr{G}, h \neq k} f_h(\mathbf{x}_i) \lambda_{hk} \qquad \forall i \in \mathscr{O}, k \in \mathscr{G} \quad (4b)$$

$$u_{ki} = \begin{cases} 1 & \text{if } k = \arg\max\{0, L_{hi} : h \in \mathscr{G}\} \\ 0 & \text{otherwise} \end{cases} \qquad \forall i \in \mathscr{O}, k \in \{0\} \cup \mathscr{G} \quad (4c)$$

$$\sum_{k \in \{0\} \cup \mathscr{G}} u_{ki} = 1 \qquad \forall i \in \mathscr{O} \quad (4d)$$

$$\sum_{i : i \in \mathscr{O}_h} u_{ki} \leq \lfloor \alpha_{hk} n_h \rfloor \qquad \forall h, k \in \mathscr{G}, h \neq k \quad (4e)$$

$$u_{ki} \in \{0, 1\} \qquad \forall i \in \mathscr{O}, k \in \{0\} \cup \mathscr{G}$$

$$L_{ki} \text{ unrestricted in sign} \qquad \forall i \in \mathscr{O}, k \in \mathscr{G}$$

$$\lambda_{hk} \geq 0 \qquad \forall h, k \in \mathscr{G}, h \neq k$$

DAMIP

$$\max \sum_{i \in \mathscr{O}} u_{y_i i}$$

$$\text{s.t. } L_{ki} = \pi_k f_k(\mathbf{x}_i) - \sum_{h \in \mathscr{G}, h \neq k} f_h(\mathbf{x}_i) \lambda_{hk} \qquad \forall i \in \mathscr{O}, k \in \mathscr{G}$$

$$t_i - L_{ki} \leq M(1 - u_{ki}) \qquad \forall i \in \mathscr{O}, k \in \mathscr{G} \quad (5a)$$

$$t_i - L_{ki} \geq \varepsilon(1 - u_{ki}) \qquad \forall i \in \mathscr{O}, k \in \mathscr{G} \quad (5b)$$

$$t_i \leq M(1 - u_{0i}) \qquad \forall i \in \mathscr{O} \quad (5c)$$

$$t_i \geq \varepsilon u_{ki} \qquad \forall i \in \mathscr{O}, k \in \mathscr{G} \quad (5d)$$

$$\sum_{k \in \{0\} \cup \mathscr{G}} u_{ki} = 1 \qquad \forall i \in \mathscr{O}$$

$$\sum_{i : i \in \mathscr{O}_h} u_{ki} \leq \lfloor \alpha_{hk} n_h \rfloor \qquad \forall h, k \in \mathscr{G}, h \neq k$$

$$u_{ki} \in \{0, 1\} \qquad\qquad \forall i \in \mathcal{O}, k \in \{0\} \cup \mathcal{G}$$

$$L_{ki} \text{ unrestricted in sign} \qquad\qquad \forall i \in \mathcal{O}, k \in \mathcal{G}$$

$$t_i \geq 0 \qquad\qquad \forall i \in \mathcal{O}$$

$$\lambda_{hk} \geq 0 \qquad\qquad \forall h, k \in \mathcal{G}, h \neq k$$

Although Anderson's model is parametric, DAMIP does not require knowledge of prior probability nor the conditional probability densities for the data. Specifically independent of the type of data, we often employ 1/G and conditional normal distribution for these two parameters respectively.

DAMIP is NP-hard for more than two groups [3, 4] and computationally it has proven to be very challenging to solve. For the work presented herein, we developed a greedy algorithm to solve the resulting DAMIP instances [31]. Given a pair of groups h and k, the greedy algorithm finds the best λ_{hk} and λ_{kh} while keeping other λ's unchanged. If we greedily find λ_{hk} and λ_{kh} for each pair of groups once, the complexity of the algorithm is $O(K^2 n \log n)$, which is very fast. Computational studies showed that the greedy algorithm can obtain high quality solutions compared to the exact solutions obtained by our optimization solver [31].

3.2   Feature Selection

To select a small set of discriminatory features, we performed feature selection using particle swarm optimization (PSO). PSO is an evolutionary computation technique to solve optimization problems, originally developed by Kennedy and Eberhart [10]. Candidate solutions of the optimization problem are represented as position vectors of particles. Let $\mathbf{x}_i$ be the position vector and $\mathbf{v}_i$ be the velocity vector of particle i. Let $\mathbf{p}_i$ be the best position vector of particle i in the history, i.e., the position possessing the best fitness value among all positions visited so far by particle i. In the initialization iteration of the PSO algorithm, $\mathbf{x}_i$ and $\mathbf{v}_i$ for each particle i are randomly generated within predetermined ranges, and they are updated in following iterations according to the formulas

$$\mathbf{v}_i \leftarrow \mathbf{v}_i \cdot \omega + (\mathbf{p}_i - \mathbf{x}_i) \cdot c_1 \cdot \text{rand}() + (\mathbf{p}_{n*(i)} - \mathbf{x}_i) \cdot c_2 \cdot \text{rand}(), \qquad (6a)$$

$$\mathbf{x}_i \leftarrow \mathbf{x}_i + \mathbf{v}_i, \qquad\qquad\qquad (6b)$$

where $n^*(i)$ is the index of the best particle (i.e., having the best objective function value in the history) in the neighborhood of the ith particle, rand() denotes a random number, and ω, c_1 and c_2 are parameters. In the velocity updating formula (6a), the terms involving the best previous position and the position of the best neighbor are considered as cognitive and social learning, respectively, of the particle. Poli et al. had a detailed overview of PSO, including the choice of parameter values in the updating equations [27].

In the feature selection application, we devised a discrete version of PSO, in which a binary vector is regarded as the position vector of a particle, indicating whether each feature is selected or not. We used the cross-validation prediction accuracy obtained by DAMIP as the objective function in PSO, and the two optimization components form the machine learning framework for our predictive model.

3.3   Cross-Validation and Blind Prediction

To obtain an unbiased estimate of the reliability and quality of the derived classification rules, tenfold cross validation is performed. In the tenfold cross validation procedure, the training set is randomly partitioned into ten subsets of roughly equal size. Ten computational experiments are then run, each of which involves a distinct training set made up of nine of the ten subsets and a test set made up of the remaining subset. The classification rule obtained via a given training set is applied to each point in the associated test set to determine to which group the rule allocates it. The process is repeated until each subset has been used once for testing. The cumulative measure of correct classification of the ten experiments provides the unbiased estimation of correct classification.

We test the predictive rules developed using blind independent data via onefold blind prediction. In onefold blind prediction, a classification rule is first developed using all the training data using the selected set of discriminatory features. This rule is then applied to each subject in the blind data to predict its group status. The percent of correct prediction of the blind data is recorded, providing a measure of overall prediction accuracy.

4   Results

The Emory data was used in two ways. First, data from one trial was used for training to develop a prediction rule, and the other trial was used to blind test. Second, data from both trials were combined, and we randomly selected 67% of the subjects in each group for tenfold cross-validation tests, and used the remaining subjects for blind prediction tests. We applied the PSO-DAMIP machine learning classification framework to identify patterns that can discriminate subjects from AD, MCI, and control groups. The best classification results as well as the selected discriminatory features in each case are shown in the following tables. In each classification result, the left part shows the counts while the right part shows the fraction. The row and column titles represent the real and predicted groups, respectively. Take Table 2 as an example. All seven AD patients are represented in row 1 (spread across the tenfold and blind prediction columns). Five of them are used for tenfold cross-validation, and two for blind prediction. For cross-validation,

Table 2 Classification results of Emory data, tenfold cross-validation and blind prediction. Five discriminatory features were selected (among the 100 features): MMSE–cMMtotal, WordList–cWL2Butter, WordList–cWL2Queen, WordList–cWL2Ticket, GDS–GDS13

Tenfold cross validation						Blind prediction						
	AD	MCI	Ctl	AD	MCI	Ctl		AD	MCI	Ctl	AD	MCI	Ctl
AD	4	1	0	0.80	0.20	0.00	AD	2	0	0	1.00	0.00	0.00
MCI	0	11	0	0.00	1.00	0.00	MCI	1	4	0	0.20	0.80	0.00
Ctl	0	0	8	0.00	0.00	1.00	Ctl	0	0	4	0.00	0.00	1.00
Unbiased estimate accuracy: 96%							Blind prediction accuracy: 91%						

Table 3 Classification results of Emory data, tenfold cross-validation and blind prediction. Four discriminatory features were selected (among the 100 features): MMSE–cMMtotal, MMSE–cMMz, WordList–cWL1Queen, GDS–GDS13

Tenfold cross validation						Blind prediction						
	AD	MCI	Ctl	AD	MCI	Ctl		AD	MCI	Ctl	AD	MCI	Ctl
AD	4	1	0	0.80	0.20	0.00	AD	2	0	0	1.00	0.00	0.00
MCI	0	10	1	0.00	0.91	0.09	MCI	0	5	0	0.00	1.00	0.00
Ctl	0	0	8	0.00	0.00	1.00	Ctl	0	1	3	0.00	0.25	0.75
Unbiased estimate accuracy: 92%							Blind prediction accuracy: 91%						

Table 4 Classification results of Emory data, tenfold cross-validation and blind prediction. Five discriminatory features were selected (among the 100 features): MMSE–cMMsRapple, WordList–cWL1Queen, WordList–cWL3Engine, GDS–GDS9, GDS–GDS13

Tenfold cross-validation						Blind prediction						
	AD	MCI	Ctl	AD	MCI	Ctl		AD	MCI	Ctl	AD	MCI	Ctl
AD	5	0	0	1.00	0.00	0.00	AD	2	0	0	1.00	0.00	0.00
MCI	0	10	1	0.00	0.91	0.09	MCI	1	4	0	0.20	0.80	0.00
Ctl	0	1	7	0.00	0.13	0.88	Ctl	0	0	4	0.00	0.00	1.00
Unbiased estimate accuracy: 92%							Blind prediction accuracy: 91%						

Table 5 Classification results of Emory data, tenfold cross-validation and blind prediction. Five discriminatory features were selected (among the 100 features): MMSE–cMMtotal, WordList–cWL3Queen, WordList–cWL2Engine, GDS–GDS13, GDS–GDS15

Tenfold cross-validation						Blind prediction						
	AD	MCI	Ctl	AD	MCI	Ctl		AD	MCI	Ctl	AD	MCI	Ctl
AD	3	2	0	0.60	0.40	0.00	AD	1	1	0	0.50	0.50	0.00
MCI	0	11	0	0.00	1.00	0.00	MCI	0	5	0	0.00	1.00	0.00
Ctl	0	0	8	0.00	0.00	1.00	Ctl	0	0	4	0.00	0.00	1.00
Unbiased estimate accuracy: 92%							Blind prediction accuracy: 91%						

4 are correctly classified (row AD, column AD) and 1 is misclassified (row AD, column MCI). The values 0.80 and 0.20 in the same positions in the next panel indicate 80% correct classification (AD to AD) and 20% misclassification (AD to MCI) for cross-validation. The value in the bottom of the tenfold cross-validation section says the overall unbiased estimate accuracy is 96%.

Table 6 Classification results of Emory data: Use Trial 1 subjects for training, and Trial 2 subjects for blind prediction. Five discriminatory features were selected (among the 100 features): MMSE–cMMsCounty, MMSE–cMMsWorld, Clock–cClockHands1, WordList–cWL2Queen, WordList–cWRyShore/WordList–cWRyArm

Training						Blind prediction					
	AD	MCI	Ctl	AD	MCI	Ctl		AD	MCI	Ctl	AD	MCI	Ctl
AD	5	0	0	1.00	0.00	0.00	AD	2	0	0	1.00	0.00	0.00
MCI	0	3	0	0.00	1.00	0.00	MCI	0	9	4	0.00	0.69	0.31
Ctl	0	0	2	0.00	0.00	1.00	Ctl	0	1	9	0.00	0.10	0.90
Accuracy: 100%							Blind prediction accuracy: 80%					

Table 7 Classification results of Emory data: Use Trial 2 subjects for training, and Trial 1 subjects for blind prediction. Five discriminatory features were selected (among the 100 features): Age, MMSE–cMMsRapple, WordList–cWL2Queen, WordList–cWL2Engine, GDS–GDS13

Training						Blind prediction					
	AD	MCI	Ctl	AD	MCI	Ctl		AD	MCI	Ctl	AD	MCI	Ctl
AD	2	0	0	1.00	0.00	0.00	AD	4	1	0	0.80	0.20	0.00
MCI	1	12	0	0.08	0.92	0.00	MCI	0	3	0	0.00	1.00	0.00
Ctl	0	0	10	0.00	0.00	1.00	Ctl	0	0	2	0.00	0.00	1.00
Accuracy: 96%							Blind prediction accuracy: 90%					

Table 8 Classification results of Emory data, tenfold cross-validation and blind prediction from 9 score-type features. Two discriminatory features were selected: MMSE–cMMtotal, Word List–cWLcorTotal

Tenfold cross-validation						Blind prediction					
	AD	MCI	Ctl	AD	MCI	Ctl		AD	MCI	Ctl	AD	MCI	Ctl
AD	4	1	0	0.80	0.20	0.00	AD	1	1	0	0.50	0.50	0.00
MCI	1	9	1	0.09	0.82	0.09	MCI	0	5	0	0.00	1.00	0.00
Ctl	0	2	6	0.00	0.25	0.75	Ctl	0	1	3	0.00	0.25	0.75
Unbiased estimate accuracy: 79%							Blind prediction accuracy: 82%					

Classification results of Emory data are shown in Tables 2–8. Tables 2–5 show the results of tenfold cross-validation and blind prediction when machine learning was performed on all 100 features and at most five discriminatory features were selected. Across all these tables, we can observe that misclassification occurs across AD and MCI, or MCI and Control. Table 6 shows the results of establishing the rule using subjects from Trial 1 and blind predicting Trial 2 subjects; and Table 7 shows the reverse. Comparing results from Tables 2–6, we observe that the predictive rule generated based solely on Trial 1 subjects is not as accurate in blind prediction when compared to the results where random mixing and selection of Trial subjects were used to establish the rule. This may be due to the fact that there are fewer subjects in Trial 1 and that discriminatory features identified there may not be as representative as those obtained for Tables 2–5. Establishing the rule using subjects from Trial 2 (Table 7) predicts well and the results are comparable to those in Tables 2–5 since the Trial size is larger, and thus the features selected are more

Table 9 Classification results of LONI/ADNI data, tenfold cross-validation and blind prediction. Five discriminatory features were selected (among the 54 features): CLOCKHAND, AVTOT5, AVTOT6, CATVEGESC, TRABERROM

Tenfold cross-validation						Blind prediction					
	AD	MCI	Ctl	AD	MCI	Ctl		AD	MCI	Ctl	AD	MCI	Ctl
AD	114	35	0	0.77	0.23	0.00	AD	56	17	1	0.76	0.23	0.01
MCI	38	175	47	0.15	0.67	0.18	MCI	21	85	22	0.16	0.66	0.17
Ctl	3	42	72	0.03	0.36	0.62	Ctl	0	22	36	0.00	0.38	0.62
Unbiased estimate accuracy: 69%						Blind prediction accuracy: 68%					

Table 10 Classification results of LONI/ADNI data, tenfold cross-validation and blind prediction. Five discriminatory features were selected (among the 54 features): AVTOT5, AVTOT6, CATVEGESC, TRABSCOR, TRABERROM

Tenfold cross-validation						Blind prediction					
	AD	MCI	Ctl	AD	MCI	Ctl		AD	MCI	Ctl	AD	MCI	Ctl
AD	113	35	1	0.76	0.23	0.01	AD	57	17	0	0.77	0.23	0.00
MCI	36	173	51	0.14	0.67	0.20	MCI	20	85	23	0.16	0.66	0.18
Ctl	1	43	73	0.01	0.37	0.62	Ctl	0	23	35	0.00	0.40	0.60
Unbiased estimate accuracy: 68%						Blind prediction accuracy: 68%					

diverse and representative. Note that the feature cWLxQueen (during which subjects are presented word lists, and must recall the word Queen) appears among all of these discriminatory sets. Further, in all analysis, there is no misclassification between AD and Control group, indicating a clear difference in the raw neuropsychological data characteristics among these two groups of patients.

Table 8 shows the results of tenfold cross-validation and blind prediction in which discriminatory patterns were selected from 9 score-type features. In this case, only two discriminatory patterns were found. The results are inferior to those obtained when features were selected from the raw data. Our study shows that classification analysis performed on the set of raw neuropsychological data yield better predictive power than those using only score-type features.

Classification results of LONI/ADNI data are shown in Tables 9 and 10. As expected, the results are far inferior to those obtained from the Emory data, due to the fact that all these features were pre-processed score-type values rather than raw data of the tests.

5 Conclusion

Systems modeling and quantitative analysis of large amounts of complex clinical and biological data may help to identify discriminatory patterns that can uncover health risks, detect early disease formation, monitor treatment and prognosis, and predict treatment outcome. In this paper, we describe a PSO-DAMIP machine-learning framework for early detection of mild cognitive impairment and Alzheimer's disease. The features used to establish the predictive rules are obtained from raw neuropsychological data. The predictive modeler and solver maximize correct classification while constraining inter-group misclassifications. The classification/predictive models (1) have the ability to classify any number of distinct groups; (2) allow incorporation of heterogeneous, and continuous/time-dependent types of attributes as input; (3) utilize a high-dimensional data transformation that minimizes noise and errors in biological and clinical data; (4) incorporate a reserved-judgment region that provides a safeguard against over-training; and (5) have successive multi-stage classification capability.

The classification results based on the neuropsychological data show that such a classification approach can be used successfully to develop a classification rule to predict AD, MCI, and normal group membership with blind prediction accuracy of over 90%. Further, our study strongly suggests that raw data of neuropsychological tests have higher potential to predict subjects from AD, MCI, and control groups than pre-processed subtotal score-like features.

The number of people affected by Alzheimer's disease is growing at a rapid rate, and the consequent increase in costs will have significant impacts on the world's economies and health care systems. Therefore, there is an urgent need to identify mechanisms that can provide early detection of the disease to allow for timely intervention. Neuropsychological tests are inexpensive, non-invasive, and can be incorporated within an annual physical examination. Thus they can serve as a baseline for early cognitive impairment or Alzheimer's disease risk prediction. The classification approach and the results discussed herein offer the potential for development of a clinical decision making tool. Further study must be conducted to validate its clinical significance and its predictive accuracy among various demographic groups and across multiple sites.

References

1. J.A. Anderson, Constrained discrimination between k populations. J. Roy. Stat. Soc. B (Methodological) **31**(1), 123–139 (1969)
2. M.W. Bondi, A.J. Jak, L. Delano-Wood, M.W. Jacobson, D.C. Delis, D.P. Salmon, Neuropsychological contributions to the early identification of Alzheimer's disease. Neuropsychol. Rev. **18**(1), 73–90 (2008)
3. J.P. Brooks, E.K. Lee, Analysis of the consistency of a mixed integer programming-based multi-category constrained discriminant model. Ann. Oper. Res. **174**(1), 147–168 (2010)
4. J.P. Brooks, E.K. Lee, Solving a mixed integer programming multi-category classification model with misclassification constraints. INFORMS J. Comput. (2011, accepted)
5. M. Brys, E. Pirraglia, K. Rich, S. Rolstad, L. Mosconi, R. Switalski, L. Glodzik-Sobanska, S. De Santi, R. Zinkowski, P. Mehta et al., Prediction and longitudinal study of CSF biomarkers in mild cognitive impairment. Neurobiol. Aging **30**(5), 682–690 (2009)
6. R. Chaves, J. Ramírez, J.M. Górriz, M. López, D. Salas-Gonzalez, I. Álvarez, F. Segovia, SVM-based computer-aided diagnosis of the Alzheimer's disease using t-test NMSE feature selection with feature correlation weighting. Neurosci. Lett. **461**(3), 293–297 (2009)

7. F.A. Feltus, E.K. Lee, J.F. Costello, C. Plass, P.M. Vertino, Predicting aberrant CpG island methylation. Proc. Natl. Acad. Sci. **100**(21), 12253–12258 (2003)
8. R.J. Gallagher, E.K. Lee, D.A. Patterson, in *An Optimization Model for Constrained Discriminant Analysis and Numerical Experiments with Iris, Thyroid, and Heart Disease Datasets.* Proceedings of the AMIA Annual Fall Symposium (American Medical Informatics Association, Bethesda, Maryland, 1996) pp. 209–213
9. R.J. Gallagher, E.K. Lee, D.A. Patterson, Constrained discriminant analysis via 0/1 mixed integer programming. Ann. Oper. Res. **74**, 65–88 (1997)
10. J. Kennedy, R. Eberhart, in *Particle Swarm Optimization.* IEEE International Conference on Neural Networks, 1995. Proceedings, vol. 4 (IEEE, NY, 1995), pp. 1942–1948
11. A. Kluger, S.H. Ferris, J. Golomb, M.S. Mittelman, B. Reisberg, Neuropsychological prediction of decline to dementia in nondemented elderly. J. Geriatric Psychiatr. Neurol. **12**(4), 168–179 (1999)
12. E.K. Lee, Large-scale optimization-based classification models in medicine and biology. Ann. Biomed. Eng. **35**(6), 1095–1109 (2007)
13. E.K. Lee, *Lecture Notes in Computer Science.* Theoretical Computer Science and General Issues. 6th International Conference, CPAIOR 2009 Pittsburgh, PA, USA, vol. 5547, pp. 27–31 (2009). ⟨http://www.springer.com/computer/theoretical+computer+science/book/978-3-642-01928-9⟩
14. E.K. Lee, T.L. Wu, *Springer Optimization and Its Applications.* Pardalos, Panos M.; Romeijn, H. Edwin (Eds.) Handbook of Optimization in Medicine, Vol. 26 pp. 1–50 (2009)
15. E.K. Lee, A.Y.C. Fung, J.P. Brooks, M. Zaider, Automated planning volume definition in soft-tissue sarcoma adjuvant brachytherapy. Phys. Med. Biol. **47**, 1891–1910 (2002)
16. E.K. Lee, R.J. Gallagher, D.A. Patterson, A linear programming approach to discriminant analysis with a reserved-judgment region. INFORMS J. Comput. **15**(1), 23–41 (2003)
17. E.K. Lee, R.J. Gallagher, A.M. Campbell, M.R. Prausnitz, Prediction of ultrasound-mediated disruption of cell membranes using machine learning techniques and statistical analysis of acoustic spectra. IEEE Trans. Biomed. Eng. **51**(1), 82–89 (2004)
18. M.M. López, J. Ramírez, J.M. Górriz, I. Álvarez, D. Salas-Gonzalez, F. Segovia, R. Chaves, SVM-based CAD system for early detection of the Alzheimer's disease using kernel PCA and LDA. Neurosci. Lett. **464**(3), 233–238 (2009)
19. O.L. Lopez, J.T. Becker, W.J. Jagust, A. Fitzpatrick, M.C. Carlson, S.T. DeKosky, J. Breitner, C.G. Lyketsos, B. Jones, C. Kawas et al., Neuropsychological characteristics of mild cognitive impairment subgroups. J. Neurol. Neurosurg. Psychiatr. **77**(2), 159–165 (2006)
20. M.T. McCabe, E.K. Lee, P.M. Vertino, A multifactorial signature of DNA sequence and polycomb binding predicts aberrant CpG island methylation. Cancer Res. **69**(1), 282–291 (2009)
21. L.K. McEvoy, C. Fennema-Notestine, J.C. Roddey, D.J. Hagler, D. Holland, D.S. Karow, C.J. Pung, J.B. Brewer, A.M. Dale, Alzheimer disease: Quantitative structural neuroimaging for detection and prediction of clinical and structural changes in mild cognitive impairment. Radiology **251**(1), 195–205 (2009)
22. C. Misra, Y. Fan, C. Davatzikos, Baseline and longitudinal patterns of brain atrophy in MCI patients, and their use in prediction of short-term conversion to AD: Results from ADNI. Neuroimage **44**(4), 1415–1422 (2009)
23. H.I. Nakaya, J. Wrammert, E.K. Lee, L. Racioppi, S. Marie-Kunze, W.N. Haining, A.R. Means, S.P. Kasturi, N. Khan, G.M. Li et al., Systems biology of vaccination for seasonal influenza in humans. Nat. Immunol. **12**(8), 786–795 (2011)
24. A.P. Nelson, M.G. O'Connor, Mild cognitive impairment: A neuropsychological perspective. CNS Spectrums **13**(1), 56–64 (2008)
25. S.E. O'Bryant, G. Xiao, R. Barber, J. Reisch, R. Doody, T. Fairchild, P. Adams, S. Waring, R. Diaz-Arrastia, A serum protein-based algorithm for the detection of Alzheimer disease. Arch. Neurol. **67**(9), 1077–1081 (2010)
26. S.E. O'Bryant, G. Xiao, R. Barber, J. Reisch, J. Hall, C.M. Cullum, R. Doody, T. Fairchild, P. Adams, K. Wilhelmsen et al., A blood-based algorithm for the detection of Alzheimer's disease. Dement. Geriatr. Cognit. Disord. **32**(1), 55–62 (2011)

27. R. Poli, J. Kennedy, T. Blackwell, Particle swarm optimization. Swarm Intell. **1**(1), 33–57 (2007)
28. T.D. Querec, R.S. Akondy, E.K. Lee, W. Cao, H.I. Nakaya, D. Teuwen, A. Pirani, K. Gernert, J. Deng, B. Marzolf et al., Systems biology approach predicts immunogenicity of the yellow fever vaccine in humans. Nat. Immunol. **10**(1), 116–125 (2008)
29. D.T. Stuss, R.L. Trites, Classification of neurological status using multiple discriminant function analysis of neuropsychological test scores. J. Consult. Clin. Psychol. **45**(1), 145 (1977)
30. M.H. Tabert, J.J. Manly, X. Liu, G.H. Pelton, S. Rosenblum, M. Jacobs, D. Zamora, M. Goodkind, K. Bell, Y. Stern, D.P. Devanand, Neuropsychological prediction of conversion to Alzheimer disease in patients with mild cognitive impairment. Arch. Gen. Psychiatr. **63**, 916–924 (2006)
31. T.L. Wu, *Classification Models for Disease Diagnosis and Outcome Analysis.* PhD thesis, Georgia Institute of Technology (2011)

Strategies for Bias Reduction in Estimation of Marginal Means with Data Missing at Random

Baojiang Chen and Richard J. Cook

Abstract Incomplete data are common in many fields of research, and interest often lies in estimating a marginal mean based on available information. This paper is concerned with the comparison of different strategies for estimating the marginal mean of a response when data are missing at random. We evaluate these methods based on the asymptotic bias, empirical bias and efficiency. We show that complete case analysis gives biased results when data are missing at random, but inverse probability weighted estimating equations (IPWEE) and a method based on the expected conditional mean (ECM) yield consistent estimators. While these methods give estimators which behave similarly in the contexts studied they are based on quite different assumptions. The IPWEE approach requires analysts to specify a model for the missing data mechanism whereas the ECM approach requires a model for the distribution of auxiliary variables driving the missing data mechanism. The latter can be a challenge in practice, particularly when the covariates are of high dimension or are a mixture of continuous and categorical variables. The IPWEE approach therefore has considerable appeal in many practical settings.

Mathematics Subject Classification (2010): Primary 62H12, Secondary 62F10

B. Chen (✉)
Department of Biostatistics, University of Nebraska Medical Center,
Omaha, NE 68198, USA
e-mail: baojiang.chen@unmc.edu

R.J. Cook
Department of Statistics and Actuarial Science, University of Waterloo,
Waterloo, ON, Canada N2L 3G1
e-mail: rjcook@uwaterloo.ca

P.M. Pardalos et al. (eds.), *Optimization and Data Analysis in Biomedical Informatics*,
Fields Institute Communications 63, DOI 10.1007/978-1-4614-4133-5_5,
© Springer Science+Business Media New York 2012

1 Introduction

In many scientific studies it is difficult to collect complete information on a response of interest. In practice, interests often lie in estimating the marginal mean of the response based on available data. For example, in assessing the effectiveness of Smoker Helpline programs, a central objective is to estimate the probability clients have quit smoking 6 months after their first use of this service. Many individuals fail to provide information on their smoking status at 6 months and so smoking status is only available for the subset of individuals who stayed in this program for this period of time. Concern then lies in the appropriate method of analysis using the available data; the implications of using standard methods of analysis depend on the nature of the missing data mechanism.

The simplest and most common way of dealing with this incomplete data is to base analyses on those individuals who provided information; this is sometimes called the *complete case analysis*. This method is valid when data are *missing completely at random* (MCAR), however when data are *missing at random* (MAR) or *missing not at random* (MNAR) the complete case analysis gives inconsistent estimators [3, 8, 15]. The MAR mechanism has been the most widely discussed in the literature, and arises when the probability a response is missing depends on observable quantities such as covariates which, while associated with the response, are not controlled for in the response model because their effects are not of scientific interest.

To deal with MAR mechanisms, a common approach is to use inverse probability weighted estimating equations (IPWEE) or inverse probability weighted generalized estimating equations (IPWGEE) in the context of longitudinal data [12, 13]. We focus on the case of a binary scalar response and hence estimating equations based on likelihood or quasi-likelihood methods [10]. As a semiparametric method, the IPWEE is robust to the distribution assumptions for the response, but the estimator may not be efficient and is generally sensitive to the misspecification of the model for the missing data process. An alternative approach is to render the missing data as ignorable by conditioning on the covariates in the response model. This ensures standard analyses are valid, but means that the analysis is not based on the desired response model since it involves conditioning on the covariates driving the missing data process. One must then compute the expectation of the conditional mean (ECM) of the response to obtain an estimate of the desired marginal mean. This ECM method will give efficient estimates of the marginal mean if (1) the effects of the covariates on the response model are correctly specified and (2) the joint distribution of the covariates is correctly specified. Misspecification of either of these aspects will yield biased estimators of the marginal mean of the response. In practice it is a challenge to model the covariates distribution, particularly when the covariates are of a high dimension with both continuous and categorical covariates, which has lead to increased interest in the use of inverse probability weights.

Little work has been done on comparing the frequency properties of estimators from these two approaches under correct model specification. We address this here and give some guidance for data analysts.

This paper is organized as follows. In Sect. 2, we introduce some methods for the estimation of the marginal mean; Sect. 3 gives some numerical studies for comparisons of different methods. In Sect. 4, we apply these methods to a skeletal metastases and a smoking prevention project studies. Concluding remarks are given in Sect. 5.

2 Estimation of the Marginal Mean

Consider a sample of n individuals yielding independent responses. Let Y_i denote the response for subject i and $X_i = (1, X_{i1}, \ldots, X_{i,p-1})'$ denote a $p \times 1$ covariate vector, $i = 1, 2, \ldots, n$. There may be a relationship between the covariates and the response which is most commonly expressed through a generalized linear model. In such cases we might write

$$g(\mu_i) = X_i'\beta, \tag{1}$$

where $\mu_i = E(Y_i|X_i;\beta)$, $g(\cdot)$ is a $1-1$ monotone differentiable link function, and β is a vector of regression coefficients. Here, however, we consider the case where interest lies simply in estimating the marginal (unconditional) mean $E(Y_i) = \mu$ in the setting where there is a dependence between X_i and Y_i (i.e. $\beta_j \neq 0$ for at least one j, $j = 1, \ldots, p-1$). We let $F(X_i;\gamma)$ denote the multivariate joint distribution of the covariate vector X_i indexed by γ and note that $\mu = \mu(\beta, \gamma) = \int E(Y_i|X_i;\beta)dF(X_i;\gamma)$.

We consider the case in which it is not possible to observe the response for all n individuals but assume that the covariates are always observed. We let R_i denote the missing data indicator for the response Y_i such that $R_i = 1$ if Y_i is observed and $R_i = 0$ if Y_i is missing; Whether the response is observed or not is governed by a stochastic model which we write in general as $P(R_i|Y_i, X_i;\alpha)$, indexed by α. If $P(R_i = 1|Y_i, X_i;\alpha)$ does in fact depend on Y_i then the data are missing not at random since whether we observe Y_i or not depends on its value. This setting is problematic for analysts since not all parameters can be identified precluding consistent estimation of μ; sensitivity analyses are recommended in this setting (cite some authors- Robbins papers) [14]. Here we focus on MAR mechanisms, for which $P(R_i = 1|Y_i, X_i;\alpha) = P(R_i = 1|X_i;\alpha)$, and hence R_i is conditionally independent of Y_i given X_i.

A naive approach for estimating μ would base analysis on available data using a standard estimating equation, given by

$$\sum_{i=1}^{n} R_i(Y_i - \mu). \tag{2}$$

This yields the estimator

$$\tilde{\mu} = \sum_{i=1}^{n} R_i Y_i / \sum_{i=1}^{n} R_i.$$

The estimating function (2) does not have expectation zero, however, and hence $\tilde{\mu}$ is not consistent for μ under a MAR mechanism. To see this consider the contribution for subject i and note that

$$E_{R,Y,X}[R_i(Y_i - \mu)] = E_X[E_{Y|X}[E_{R|Y,X}[R_i(Y_i - \mu); \alpha]; \beta]; \gamma]$$
$$= E_X[E_{Y|X}[P(R_i = 1|X_i; \alpha)(Y_i - \mu); \beta]; \gamma]$$
$$= E_X[P(R_i = 1|X_i; \alpha)(E(Y_i|X_i; \beta) - \mu)]; \gamma]$$
$$\neq 0.$$

Only under a MCAR mechanism (i.e. if $P(R_i = 1|X_i; \alpha) = P(R_i = 1; \alpha)$) can $P(R_i = 1|X_i; \alpha) = P(R_i = 1; \alpha)$ be factored out of the final expectation with respect to the covariate, giving an unbiased estimating function. In the following we introduce two alternative methods to this naive complete case analysis, each of which give consistent estimators for the marginal mean of the response, μ, under the MAR mechanism.

2.1 Expected Conditional Mean

For this approach, we first estimate the conditional mean of the response given the covariates using the available data, and then estimate the marginal mean by

$$\mu = E(Y_i; \beta, \gamma) = E_X[E_{Y|X}(Y_i|X_i; \beta); \gamma].$$

To estimate the conditional mean $E(Y_i|X_i; \beta)$, we generally consider model (1) presuming the covariate effects are adequately modeled, and estimate the parameter β. Often we employ the following estimating equation

$$U(\beta) = \sum_{i=1}^{n} U_i(\beta) = 0, \tag{3}$$

where $U_i(\beta) = R_i D_i V_i^{-1}(Y_i - \mu_i)$ with $D_i = \partial \mu_i / \partial \beta$, and $V_i = \text{var}(Y_i|X_i)$ only depends on the conditional mean μ_i, as commonly adopted for models in the exponential family or under quasi-likelihood. Solving (3) gives a consistent estimator for β if the conditional mean μ_i is correctly specified, since

$$E_{R,Y,X}[U_i(\beta)] = E_{R,Y,X}[R_i D_i V_i^{-1}(Y_i - \mu_i); \alpha, \beta, \gamma]$$

$$= E_{Y,X}[P(R_i = 1|X_i;\alpha) \cdot D_i V_i^{-1}(Y_i - \mu_i); \beta, \gamma]$$
$$= E_X[P(R_i = 1|X_i;\alpha) D_i V_i^{-1} \cdot E_{Y|X}(Y_i - \mu_i; \beta); \gamma]$$
$$= 0.$$

This implies that under a MAR mechanism, we can obtain a consistent estimator for the conditional mean μ_i based on a complete case analysis, if the model (1) is correctly specified. While this is appealing, our goal ultimately, is to estimate the marginal mean of the response, μ. We can achieve this by subsequently taking the expectation with respect to the covariate distribution. Specifically, we write this as $\widehat{\mu} = \int_X E(Y_i|X_i; \widehat{\beta}) dF(X, \widehat{\gamma})$. In practice this is carried out by summing $E(Y_i|X_i; \widehat{\beta}) dF(X, \widehat{\gamma})$ over all possible realizations of the covariate vector via

$$\sum_{x \in \Omega_X} E(Y|X = x; \widehat{\beta}) P(X = x; \widehat{\gamma})$$

where Ω_X is the covariate sample space and $P(X = x; \widehat{\gamma})$ is the estimated probability $X = x$ under $F(X, \widehat{\gamma})$. Note that if X is discrete, then this can be estimated nonparametrically based on empirical frequencies. The challenge arises when one or more elements of X are continuous. The difficulties in estimating such a distribution nonparametrically is sometimes referred to as the "curse of dimensionality" [1].

Let $\phi = (\beta', \gamma')'$ and $\text{cov}(\widehat{\phi})$ denote the covariance matrix for $\widehat{\phi}$. The variance of $\widehat{\mu}$ can be estimated using the delta method, which gives

$$\text{var}(\widehat{\mu}) = \frac{\partial \widehat{\mu}(\beta)}{\partial \beta'} \left[\text{cov}(\widehat{\beta}) \right] \frac{\partial \widehat{\mu}(\beta)}{\partial \beta} \Big|_{\beta = \widehat{\beta}},$$

when γ (the covariate distribution) is known, and

$$\text{var}(\widehat{\mu}) = \frac{\partial \widehat{\mu}(\phi)}{\partial \phi'} \left[\text{cov}(\widehat{\phi}) \right] \frac{\partial \widehat{\mu}(\phi)}{\partial \phi} \Big|_{\phi = \widehat{\phi}},$$

when γ is estimated.

To summarize, the ECM approach disregards all data with missing values, but through conditioning on the covariate process, renders missingness unimportant for estimation of β and hence μ_i. Under this approach, however, in order to get a consistent estimator of μ, as mentioned earlier, it requires not only the specification of the response model in (1), but also the specification of the joint distribution of covariates. Specification of the covariates distribution is a challenge in practice, especially when covariates are of high dimension, or when there are both continuous and categorical covariates. In the following, we describe an approach to direct estimators of the marginal mean based on inverse weighting; this approach does not require specification of the covariate distribution.

2.2 Inverse Probability Weighted Estimating Equations

Following [12], we specify the weighted estimating equation

$$U^*(\mu, \alpha) = \sum_{i=1}^{n} U_i^*(\mu, \alpha) = 0 \tag{4}$$

for $\widehat{\mu} = E(Y_i)$, where $U_i^*(\mu, \alpha) = R_i / \pi_i(\alpha)(Y_i - \mu)$, and $\pi_i(\alpha) = P(R_i = 1 | Z_i; \alpha)$; we write Z_i here since there may be covariates in addition to those in X_i which affect drop-out, so we consider $Z_i = (X_i', V_i')'$ where V_i represents these possible additional covariates which are conditionally independent of Y_i given X_i. The key point is that, since Z_i contains X_i, R_i and Y_i are conditionally independent given Z_i. The estimating equation above gives a consistent estimate for the marginal mean $\mu = E(Y_i)$ if the missing data model is correctly specified, since

$$\begin{aligned} E_{R,Y,Z}[U_i^*(\mu, \alpha)] &= E_{Y,Z}\left\{ E_{R|Y,Z}\left[\frac{R_i}{\pi(Z_i; \alpha)}(Y_i - \mu); \alpha \right]; \beta, \gamma \right\} \\ &= E_{Y,Z}[Y_i - \mu; \beta, \gamma] \\ &= E_X[E_{Y|X}(Y_i - \mu; \beta); \gamma] \\ &= 0. \end{aligned}$$

If $\pi(Z_i; \alpha)$ is known, we can estimate μ by solving (4), to get a Horvitz-Thompson [5] estimator of the form

$$\widehat{\mu} = \frac{\sum_{i=1}^{n} R_i Y_i / \pi(Z_i; \alpha)}{\sum_{i=1}^{n} R_i / \pi(Z_i; \alpha)}. \tag{5}$$

In practice, $\pi(Z_i; \alpha)$ is unknown and has to be estimated consistently. We can then replace $\pi(Z_i; \alpha)$ by a consistent estimator $\widehat{\pi}_i$ in (5) to give the estimator $\widehat{\mu}$, which will still be consistent. Therefore, using the inverse weighting approach, we can obtain a consistent estimator of the marginal mean if we correctly specify the missing data model.

For the missing indicator R_i, we often build a generalized linear model for the conditional mean $\pi_i = \pi(Z_i; \alpha) = P(R_i = 1 | Z_i; \alpha)$ via a logistic link, say, as follows

$$\text{logit } \pi_i = Z_i'\alpha,$$

where Z_i is a covariate vector reflecting the missingness, and α is the corresponding coefficient vector. As can be seen from above, here we require Z_i to capture a sufficient amount of information so that $R_i \perp Y_i | Z_i$. We can use maximum likelihood to estimate α by maximizing

$$L = \prod_{i=1}^{n} L_i(\alpha),$$

where $L_i(\alpha) = \pi_i^{r_i}(1 - \pi_i)^{1-r_i}$ or equivalently solving $S(\alpha) = \sum_{i=1}^{n} S_i(\alpha) = 0$ where $S_i(\alpha) = \partial \log L_i / \partial \alpha'$ is the contribution to the score vector for α from subject i, $i = 1, \ldots, n$.

For the estimation of the variance of $\widehat{\mu}$, we can employ the method of Robins et al. [12]. Under the regularity conditions given in the appendix of [12], and when α is known as, say, α^*, we have

$$n^{1/2}\left(\widehat{\mu} - \mu_0\right) \to N\left(0, \Gamma^{-1}(\mu_0, \alpha^*)\Sigma^*\left[\Gamma^{-1}\left(\mu_0, \alpha^*\right)\right]'\right),$$

where μ_0 is the true value of μ,

$$\Gamma\left(\mu_0, \alpha^*\right) = E\left[\partial U_i^*(\mu_0, \alpha^*)/\partial \mu\right],$$

and

$$\Sigma^* = E\left[U_i^*(\mu_0, \alpha^*)\left[U_i^*(\mu_0, \alpha^*)\right]'\right].$$

The variance of $\widehat{\mu}$ can be estimated by

$$\widehat{\Gamma}^{-1}(\widehat{\mu}, \alpha^*)\widehat{\Sigma}(\widehat{\mu}, \alpha^*)\left[\widehat{\Gamma}^{-1}(\widehat{\mu}, \alpha^*)\right]'$$

with

$$\widehat{\Gamma}(\widehat{\mu}, \alpha^*) = n^{-1}\sum_{i=1}^{n} \partial U_i^*(\widehat{\mu}, \alpha^*)/\partial \mu,$$

and

$$\widehat{\Sigma}(\widehat{\mu}, \alpha^*) = n^{-1}\sum_{i=1}^{n}\left[U_i^*\left(\widehat{\mu}, \alpha^*\right) U_i^{*'}\left(\widehat{\mu}, \alpha^*\right)\right].$$

Typically α is unknown and must be estimated, in which case the variability in the estimate of α must be addressed. Under regularity conditions of [12], and under the assumption that the missing data model is correctly specified, we have

$$n^{1/2}(\widehat{\mu} - \mu_0) \to N(0, \Gamma^{-1}(\mu_0, \alpha_0)\Sigma(\mu_0, \alpha_0)[\Gamma^{-1}(\mu_0, \alpha_0)]'),$$

where α_0 is the true value of α, $\Gamma(\mu_0, \alpha_0) = E[\partial U_i^*(\mu_0, \alpha_0)/\partial \mu]$, and $\Sigma(\mu_0, \alpha_0) = E[Q_i(\mu_0, \alpha_0)Q_i'(\mu_0, \alpha_0)]$ with

$$Q_i(\mu_0, \alpha_0) = U_i^*(\mu_0, \alpha_0) - E[\partial U_i^*(\mu_0, \alpha_0)/\partial \alpha']\{E[\partial S_i(\alpha_0)/\partial \alpha']\}^{-1} S_i(\alpha_0).$$

A brief sketch of the proof follows.

Note that $n^{-1/2}\sum_{i=1}^{n} U_i^*(\widehat{\mu}, \widehat{\alpha}) = 0$, and based on a Taylor series expansion in the neighborhood of (μ_0, α_0), we have

$$0 = n^{-1/2} \sum_{i=1}^{n} U_i^*(\mu_0, \alpha_0) + \Gamma(\mu_0, \alpha_0) n^{1/2} (\widehat{\mu} - \mu_0)$$

$$+ E\left[\partial U_i^*(\mu, \alpha_0)/\partial \alpha'\right] n^{1/2} (\widehat{\alpha} - \alpha_0) + o_p(1). \tag{6}$$

Similarly, since $n^{-1/2} \sum_{i=1}^{n} S_i(\widehat{\alpha}) = 0$, and based on the Taylor expansion in the neighborhood of α_0 and following some algebraic manipulations, we have

$$n^{1/2}(\widehat{\alpha} - \alpha_0) = -\left\{E\left[\partial S_i(\alpha_0)/\partial \alpha'\right]\right\}^{-1} n^{-1/2} \sum_{i=1}^{n} S_i(\alpha_0) + o_p(1). \tag{7}$$

If we plug (7) into (6), we obtain

$$0 = n^{-1/2} \sum_{i=1}^{n} U_i^*(\mu_0, \alpha_0) + \Gamma(\mu_0, \alpha_0) n^{1/2} (\widehat{\mu} - \mu_0)$$

$$- E[\partial U_i^*(\mu_0, \alpha_0)/\partial \alpha']\{E[\partial S_i(\alpha_0)/\partial \alpha']\}^{-1} n^{-1/2} \sum_{i=1}^{n} S_i(\alpha_0) + o_p(1).$$

If $\Gamma(\mu_0, \alpha_0)$ is nonsingular, we have

$$n^{1/2}(\widehat{\mu} - \mu_0) = -\Gamma^{-1}(\mu_0, \alpha_0) n^{-1/2} \sum_{i=1}^{n} Q_i(\mu_0, \alpha_0) + o_p(1),$$

and the asymptotic distribution of $n^{1/2}(\widehat{\mu} - \mu_0)$ follows by Slutsky's theorem and the central limit theorem.

To conduct inference regarding μ, the variance of $\widehat{\mu}$ can be estimated by

$$\widehat{\Gamma}^{-1}(\widehat{\mu}, \widehat{\alpha}) \widehat{\Sigma}(\widehat{\mu}, \widehat{\alpha}) \left[\widehat{\Gamma}^{-1}(\widehat{\mu}, \widehat{\alpha})\right]'$$

with

$$\widehat{\Gamma}(\widehat{\mu}, \widehat{\alpha}) = n^{-1} \sum_{i=1}^{n} \partial U_i^*(\widehat{\mu}, \widehat{\alpha})/\partial \mu,$$

$$\widehat{\Sigma}(\widehat{\mu}, \widehat{\alpha}) = n^{-1} \sum_{i=1}^{n} \widehat{Q}_i(\widehat{\mu}, \widehat{\alpha}) \widehat{Q}_i'(\widehat{\mu}, \widehat{\alpha}),$$

and

$$\widehat{Q}_i(\widehat{\mu}, \widehat{\alpha}) = U_i^*(\widehat{\mu}, \widehat{\alpha}) - \left[n^{-1} \sum_{i=1}^{n} \partial U_i^*(\widehat{\mu}, \widehat{\alpha})/\partial \alpha'\right] \left[n^{-1} \sum_{i=1}^{n} \partial S_i(\widehat{\alpha})/\partial \alpha'\right]^{-1} S_i(\widehat{\alpha}).$$

The appeal of this approach is that we do not need to specify the distributions of the covariates; furthermore, this method is robust to the misspecification of the variance function for the response model since we can view the associated estimating equation as a quasi-score equation. In practice, the missing data indicator is relatively easy to model and diagnostic checks are available for missing response models. As a semiparametric method, the estimate may not be efficient [7].

3 An Empirical Study of Finite Sample Bias and Efficiency

Here we perform a simulation study to investigate the frequency properties of the methods discussed in the previous section. We assume that there are two binary covariates X_1 and X_2 with $E(X_1) = p_1 = 0.5$, $E(X_2) = p_2 = 0.5$, and the correlation coefficient is ρ. This gives $\pi_{11} = P(X_1 = 1, X_2 = 1) = p_1 p_2 + \rho[p_1(1-p_1)p_2(1-p_2)]^{1/2}$, $\pi_{10} = P(X_1 = 1, X_2 = 0) = p_1 - \pi_{11}$, $\pi_{01} = P(X_1 = 0, X_2 = 1) = p_2 - \pi_{11}$, and $\pi_{00} = P(X_1 = 0, X_2 = 0) = 1 - p_1 - p_2 + \pi_{11}$. We further assume the response Y is binary, and the model for $\mu_i = E(Y_i | X_{i1}, X_{i2})$ is

$$\text{logit } \mu_i = \beta_0 + \beta_1 X_{i1} + \beta_2 X_{i2}.$$

The true values are $\beta_0 = \log(1.5)$, $\beta_1 = \log(2)$, and $\beta_2 = \log(2)$.
For the missing data model, we assume

$$\text{logit } \pi_i = \alpha_0 + \alpha_1 X_{i1} + \alpha_2 X_{i2},$$

and set $\alpha_0 = -1$. Here, we assume $\alpha_1 = \alpha_2 = \alpha$ and vary it from 1 to 4 to study the performance of the estimates as a function of the strength of the MAR mechanism.

We consider five methods of analysis that are routinely used in practice. The first method is the naive method based on estimating equation (2). The second method, called IPWEE1, is based on the IPW estimating equation using the true weights π_i. The third method, called IPWEE2, is based on the IPW estimating equation using the estimated weights $\hat{\pi}_i$. The fourth method, called ECM1, is the ECM method using the true joint distribution of X_1 and X_2 and the fifth method (ECM2) is the ECM method using the empirical distribution of X_1 and X_2, which is given by $\hat{\pi}_{11} = n_{11}/n$, $\hat{\pi}_{10} = n_{10}/n$, $\hat{\pi}_{01} = n_{01}/n$, and $\hat{\pi}_{00} = n_{00}/n$, where n_{jk} is the number of subjects with $X_1 = j$ and $X_2 = k$ for $j, k = 0, 1$.

Tables 1–4 report the results for the five methods, where BIAS is the percent relative bias, ASE is the average standard error, ESE is the empirical standard error, and ECP is the empirical coverage probability (%) for the nominal 95% level. The naive method gives biased estimates of the marginal mean; as the proportion of missing observation increases, the biases increase; also as the correlation between the two covariates (ρ) increases, the biases increase. The other four methods give negligible biases and good coverage probabilities in all settings. The IPWEE1 and IPWEE2 methods give very similar empirical standard errors when the missing

Table 1 Empirical performance of five methods for the estimation of marginal mean of the response: IPWEE1 is the IPWEE method using the true weight; IPWEE2 is the IPWEE method using the estimated weight; ECM1 is the ECM method using the true distribution of the covariates; and ECM2 is the ECM method using the empirical distribution of the covariates (missing proportion is about 50%, and the number of simulations is 2,000)

Method	n = 200				n = 500				n = 1,000			
	BIAS	ASE	ESE	ECP	BIAS	ASE	ESE	ECP	BIAS	ASE	ESE	ECP
$\rho = 0$												
Naive	5.6	0.042	0.041	80.9	5.2	0.027	0.026	67.8	5.4	0.019	0.020	44.2
IPWEE1	1.6	0.048	0.048	93.6	1.2	0.031	0.031	93.5	1.3	0.022	0.023	94.4
IPWEE2	1.6	0.048	0.048	93.2	1.2	0.030	0.030	93.5	1.3	0.023	0.023	94.2
ECM1	1.5	0.047	0.047	93.5	1.3	0.030	0.030	93.6	1.3	0.022	0.022	94.4
ECM2	1.6	0.047	0.047	93.2	1.2	0.030	0.030	93.6	1.3	0.023	0.023	94.0
$\rho = 0.3$												
Naive	6.2	0.041	0.041	77.9	6.1	0.026	0.027	57.4	6.2	0.019	0.020	31.4
IPWEE1	0.9	0.050	0.049	94.0	0.9	0.032	0.032	93.1	1.0	0.022	0.023	94.2
IPWEE2	0.9	0.049	0.049	93.1	0.9	0.031	0.031	93.3	1.0	0.022	0.022	94.2
ECM1	0.9	0.048	0.049	92.9	0.9	0.031	0.031	93.5	1.0	0.022	0.022	94.0
ECM2	0.9	0.048	0.049	92.9	0.9	0.031	0.031	93.6	1.1	0.022	0.022	93.8
$\rho = 0.6$												
Naive	7.0	0.041	0.041	73.2	6.9	0.026	0.027	51.0	6.8	0.019	0.018	22.4
IPWEE1	0.6	0.051	0.052	93.8	0.4	0.033	0.033	93.9	0.4	0.023	0.022	96.0
IPWEE2	0.5	0.050	0.052	93.3	0.4	0.032	0.032	93.9	0.3	0.023	0.022	95.2
ECM1	0.5	0.049	0.051	93.1	0.4	0.031	0.032	93.5	0.3	0.022	0.022	95.6
ECM2	0.6	0.049	0.051	92.8	0.4	0.031	0.032	93.4	0.4	0.022	0.022	95.4
$\rho = 0.9$												
Naive	7.7	0.041	0.042	68.0	7.6	0.026	0.025	41.1	7.7	0.018	0.018	15.6
IPWEE1	0.1	0.052	0.055	93.1	-0.0	0.033	0.033	94.8	0.0	0.024	0.023	95.6
IPWEE2	0.0	0.051	0.055	93.3	-0.0	0.033	0.033	94.4	0.0	0.023	0.022	95.4
ECM1	-0.0	0.050	0.054	92.4	-0.0	0.032	0.032	94.7	0.0	0.023	0.022	95.0
ECM2	0.0	0.050	0.054	92.9	-0.0	0.032	0.033	94.1	0.0	0.023	0.022	95.2

Table 2 Empirical performance of five methods for the estimation of marginal mean of the response: IPWEE1 is the IPWEE method using the true weight; IPWEE2 is the IPWEE method using the estimated weight; ECM1 is the ECM method using the true distribution of the covariates; and ECM2 is the ECM method using the empirical distribution of the covariates (missing proportion is about 35%, and the number of simulations is 2,000)

Method	n = 200				n = 500				n = 1,000			
	BIAS	ASE	ESE	ECP	BIAS	ASE	ESE	ECP	BIAS	ASE	ESE	ECP
$\rho = 0$												
Naive	6.0	0.036	0.037	73.3	5.9	0.023	0.023	52.7	5.9	0.016	0.016	20.8
IPWEE1	1.4	0.044	0.046	92.1	1.4	0.028	0.029	93.4	1.3	0.020	0.019	94.0
IPWEE2	1.4	0.044	0.046	92.7	1.3	0.028	0.029	93.3	1.3	0.020	0.019	94.0
ECM1	1.4	0.042	0.043	92.1	1.3	0.027	0.027	93.8	1.3	0.019	0.019	93.8
ECM2	1.4	0.042	0.044	91.6	1.3	0.027	0.028	93.7	1.3	0.019	0.019	93.8
$\rho = 0.3$												
Naive	7.1	0.036	0.035	67.9	7.1	0.023	0.023	38.4	7.1	0.016	0.016	9.2
IPWEE1	1.0	0.047	0.046	93.6	1.0	0.030	0.030	93.1	1.2	0.021	0.021	94.0
IPWEE2	0.9	0.046	0.046	93.6	1.0	0.029	0.029	94.1	1.1	0.021	0.021	94.0
ECM1	0.9	0.044	0.044	93.8	1.0	0.028	0.028	94.1	1.1	0.020	0.020	93.8
ECM2	0.9	0.044	0.045	93.7	1.0	0.028	0.029	93.7	1.1	0.020	0.020	93.7
$\rho = 0.6$												
Naive	8.3	0.036	0.035	58.3	8.1	0.023	0.023	28.2	8.0	0.016	0.016	5.4
IPWEE1	0.8	0.049	0.047	94.1	0.4	0.031	0.032	93.5	0.2	0.022	0.022	95.6
IPWEE2	0.8	0.048	0.047	94.2	0.4	0.031	0.031	93.7	0.2	0.022	0.022	94.8
ECM1	0.9	0.047	0.046	94.2	0.5	0.030	0.030	93.5	0.2	0.021	0.021	94.8
ECM2	0.9	0.046	0.046	93.9	0.5	0.030	0.031	93.5	0.2	0.021	0.021	94.0
$\rho = 0.9$												
Naive	9.4	0.036	0.036	52.0	9.1	0.023	0.023	18.4	9.4	0.016	0.016	1.4
IPWEE1	0.3	0.051	0.051	93.7	−0.2	0.033	0.032	94.0	0.1	0.023	0.023	94.8
IPWEE2	0.2	0.050	0.051	93.7	−0.3	0.032	0.032	94.3	0.1	0.023	0.023	94.6
ECM1	0.1	0.049	0.050	93.6	−0.3	0.031	0.031	93.9	0.1	0.022	0.021	94.4
ECM2	0.3	0.049	0.051	92.8	−0.2	0.031	0.032	94.4	0.1	0.022	0.023	94.6

Table 3 Empirical performance of five methods for the estimation of marginal mean of the response: IPWEE1 is the IPWEE method using the true weight; IPWEE2 is the IPWEE method using the estimated weight; ECM1 is the ECM method using the true distribution of the covariates; and ECM2 is the ECM method using the empirical distribution of the covariates (missing proportion is about 27%, and the number of simulations is 2,000)

Method	n = 200				n = 500				n = 1,000			
	BIAS	ASE	ESE	ECP	BIAS	ASE	ESE	ECP	BIAS	ASE	ESE	ECP
ρ = 0												
Naive	5.6	0.034	0.035	74.0	5.8	0.022	0.021	50.6	5.8	0.015	0.015	20.8
IPWEE1	1.1	0.043	0.045	92.1	1.3	0.027	0.027	93.2	1.3	0.020	0.018	94.0
IPWEE2	1.0	0.043	0.045	92.0	1.3	0.027	0.027	93.5	1.3	0.019	0.018	94.0
ECM1	1.0	0.040	0.041	93.0	1.3	0.025	0.025	94.0	1.4	0.018	0.017	94.0
ECM2	1.0	0.040	0.042	92.8	1.3	0.025	0.025	93.7	1.3	0.018	0.017	93.8
ρ = 0.3												
Naive	6.8	0.034	0.035	68.1	7.0	0.022	0.021	36.0	6.7	0.015	0.016	11.8
IPWEE1	0.8	0.046	0.047	92.1	0.9	0.029	0.029	93.7	0.8	0.021	0.021	94.2
IPWEE2	0.8	0.045	0.047	92.4	0.9	0.029	0.029	93.9	0.7	0.020	0.021	94.2
ECM1	0.8	0.042	0.043	93.1	0.9	0.027	0.027	93.7	0.7	0.019	0.020	94.2
ECM2	0.8	0.042	0.043	92.8	0.9	0.027	0.027	93.7	0.7	0.019	0.020	94.2
ρ = 0.6												
Naive	8.0	0.035	0.035	59.7	8.1	0.022	0.023	24.7	8.2	0.016	0.015	3.2
IPWEE1	0.5	0.048	0.048	94.2	0.6	0.031	0.031	93.2	0.6	0.022	0.022	94.2
IPWEE2	0.4	0.048	0.048	93.9	0.6	0.030	0.031	94.6	0.6	0.021	0.022	94.2
ECM1	0.4	0.045	0.046	93.3	0.6	0.029	0.029	94.4	0.5	0.020	0.020	94.6
ECM2	0.5	0.045	0.046	92.8	0.6	0.029	0.030	93.8	0.5	0.020	0.020	95.2
ρ = 0.9												
Naive	9.6	0.035	0.035	46.9	9.4	0.022	0.023	15.2	9.4	0.016	0.016	1.6
IPWEE1	0.4	0.051	0.049	94.2	-0.0	0.032	0.033	93.4	0.1	0.023	0.024	94.4
IPWEE2	0.3	0.050	0.049	94.4	-0.1	0.032	0.033	93.3	0.1	0.022	0.024	94.2
ECM1	0.3	0.049	0.048	93.9	-0.1	0.031	0.033	93.2	0.1	0.022	0.023	94.2
ECM2	0.4	0.049	0.049	93.7	-0.1	0.031	0.033	93.9	0.1	0.022	0.024	93.8

Table 4 Empirical performance of five methods for the estimation of marginal mean of the response: IPWEE1 is the IPWEE method using the true weight; IPWEE2 is the IPWEE method using the estimated weight; ECM1 is the ECM method using the true distribution of the covariates; and ECM2 is the ECM method using the empirical distribution of the covariates (missing proportion is about 20%, and the number of simulations is 2,000)

Method	$n = 200$				$n = 500$				$n = 1{,}000$			
	BIAS	ASE	ESE	ECP	BIAS	ASE	ESE	ECP	BIAS	ASE	ESE	ECP
$\rho = 0$												
Naive	5.6	0.033	0.032	74.4	5.6	0.021	0.021	51.2	5.7	0.015	0.014	18.2
IPWEE1	1.3	0.043	0.042	92.8	1.2	0.027	0.028	93.7	1.3	0.019	0.019	93.8
IPWEE2	1.3	0.042	0.042	92.6	1.2	0.027	0.027	93.5	1.3	0.019	0.018	94.0
ECM1	1.2	0.039	0.038	93.7	1.2	0.025	0.025	93.8	1.2	0.017	0.017	94.2
ECM2	1.2	0.039	0.038	93.3	1.2	0.025	0.025	93.7	1.2	0.017	0.017	94.0
$\rho = 0.3$												
Naive	6.7	0.034	0.033	67.4	6.9	0.021	0.022	36.8	6.7	0.015	0.015	10.2
IPWEE1	0.9	0.045	0.046	92.8	1.0	0.029	0.029	94.0	0.8	0.021	0.021	94.2
IPWEE2	0.8	0.045	0.046	92.6	1.0	0.029	0.028	94.0	0.8	0.020	0.021	93.8
ECM1	0.8	0.042	0.042	93.8	0.9	0.026	0.027	94.1	0.8	0.019	0.019	94.6
ECM2	0.8	0.041	0.042	93.2	0.9	0.026	0.027	93.9	0.9	0.019	0.019	94.1
$\rho = 0.6$												
Naive	7.9	0.034	0.034	59.2	8.0	0.022	0.022	24.5	8.0	0.015	0.015	4.0
IPWEE1	0.4	0.048	0.048	93.9	0.5	0.031	0.031	95.0	0.5	0.022	0.022	94.2
IPWEE2	0.3	0.048	0.048	93.7	0.4	0.030	0.031	94.2	0.5	0.021	0.021	94.6
ECM1	0.4	0.045	0.045	94.1	0.4	0.028	0.029	94.3	0.6	0.020	0.020	94.2
ECM2	0.5	0.045	0.045	94.0	0.4	0.028	0.029	94.0	0.6	0.020	0.020	94.2
$\rho = 0.9$												
Naive	9.4	0.035	0.036	49.3	9.4	0.022	0.022	15.2	9.5	0.016	0.015	0.6
IPWEE1	0.1	0.051	0.052	92.9	0.0	0.032	0.032	95.2	0.1	0.023	0.022	95.8
IPWEE2	0.0	0.050	0.052	92.7	−0.0	0.032	0.032	95.0	0.0	0.023	0.022	96.0
ECM1	0.1	0.049	0.051	92.4	0.1	0.031	0.031	95.4	0.1	0.022	0.021	95.8
ECM2	0.1	0.048	0.051	92.4	0.0	0.031	0.031	94.8	0.0	0.022	0.022	95.8

proportion is small; when the missing proportion increases, the IPWEE1 method seems to be a little bit less efficient than the IPWEE2 method. The ECM1 and ECM2 methods give very similar empirical standard errors for the different settings. As the missing proportion decreases, IPWEE2 gives bigger standard errors than ECM2, indicating that the ECM2 estimator can be more efficient than the IPWEE2 estimator but this gain is negligible and seems to depend somewhat on the correlation coefficient ρ, the missing proportion, and the sample size.

4 Applications

4.1 Application to a Study of Patients with Skeletal Metastases

In this subsection, we apply the proposed methods to a bone metastases data set [4]. Women with advanced breast cancer often experience bone metastases. From January 1991 to March 1994, the Protocol 19 Aredia Breast Cancer Study Group of Novartis Pharmaceuticals Inc. conducted a randomized clinical trial at 97 sites in the United States, Canada, Australia and New Zealand. The osteoclast activating factors released by tumor cells cause destruction of bone, which in turn leads to the occurrence of the aforementioned skeletal complications. Radiographic surveys of bone lesions were performed and new bone lesions were recorded. Covariates of interests include age at study entry (coded as AGE: 1 for age $\geq$ 50, 0 for age < 50), ECOG score at study entry (coded as ECOG: 1 for two or more, 0 otherwise), the number of fractures at baseline (coded as FRACT: 1 for one or more, 0 for none), pain score at study entry (coded as PSCORE) which is coded as four levels based on the 25, 50 and 75% quantiles. Two hundred and twenty patients entered the study and were intended to be assessed at baseline, 6 months and 12 months from the baseline. Here we are interested in the proportion of the subjects who experienced a new bone lesions after 12 months from the baseline. The response defined here is the indicator for a new lesion at the 12 month from the base line. However, the collected measurements are incomplete. The missing proportion for the lesions is 25% for patients at 12 months.

The results for the marginal mean are listed in Table 5. Note that there is little difference among the three methods since all the covariates are not significant in the missing data model, which indicates that the missing completely at random (MCAR) may be appropriate here. See Table 6.

Table 5 Results of estimation for the marginal mean for a bone metastases study data

Parameter	Estimator	SE	95%CI
μ_{naive}	0.3576	0.0373	(0.2844, 0.4307)
μ_{IPWEE}	0.3612	0.0375	(0.2877, 0.4346)
μ_{ECM}	0.3612	0.0372	(0.2883, 0.4341)

Table 6 Missing data model for a bone metastases study data

Parameter	Estimator	SE	p-value
INTERCEPT	1.036	0.394	0.009
AGE	0.338	0.325	0.298
ECOG	0.125	0.382	0.743
FRACT	0.385	0.437	0.377
PSCORE1	−0.492	0.487	0.312
PSCORE2	−0.517	0.465	0.912
PSCORE3	−0.432	0.429	0.313

Table 7 Missing data model for the smoking prevention project data

Parameter	Estimator	SE	p-value
INTERCEPT	2.484	0.132	<0.001
TRT	0.054	0.115	0.640
GENDER	−0.309	0.097	0.001
SMR	−0.481	0.100	<0.001

4.2 Application to a Smoking Prevention Project

The Waterloo Smoking Prevention Project (WSPP) is a randomized longitudinal study designed to investigate smoking behavior among school children [2]. We report here on the results of some analysis of data from WSPP4, the fourth study in the series. A total of 100 schools in seven Ontario school boards were randomized to dispense either the regular health education programmes provided by the school or a more intensive anti-smoking programme delivered by either a specially trained teacher or a public health nurse. Questionnaires regarding smoking attitudes and behavior were administered annually from grade 6 to grade 8. One of the aims of this study is to investigate the proportion of children who smoke in grade 8.

The smoking status based on the responses to the questionnaire items can be represented by a binary variable: $Y_i = 1$ indicates subject i is a smoker in grade 8, and 0 otherwise. Along with the responses, the factors that may influence the children's smoking behavior were recorded. These covariates include gender (coded as GENDER, 0–female, 1–male), treatment effect (coded as TRT, 0–control; 1–intervention), social models risk score (coded as SMR, 0–none of parents, siblings or friends smoke; 1–at least one of parents, siblings or friends smoke). There are 4,409 subjects in the data set who are present at grade 8 with fully observed covariates. About 11.14% subjects have missing observations.

The missing data model is listed in Table 7. The significance of GENDER and SMR demonstrates that the missing at random assumption may be appropriate here for the missing mechanism. The results for the marginal mean are listed in Table 8. It is seen that the IPW and ECM methods give very similar results, which are a little bit different from the naive method.

Parameter	Estimator	SE	95%CI
μ_{naive}	0.1988	0.0064	(0.1863, 0.2113)
μ_{IPWEE}	0.2005	0.0064	(0.1879, 0.2131)
μ_{ECM}	0.2005	0.0063	(0.1881, 0.2129)

Table 8 Results of estimation for the marginal mean for a smoking prevention project data

5 Discussion

There is no bias in estimating the marginal mean by averaging the observed responses when data are MCAR. However, under a MAR mechanism, such a naive analysis generally gives biased results, and this bias depends on the association between the covariates and the association between the covariates and the missing indicator. We described here two methods (ECM and IPWEE) which can be used to address this bias. The ECM is appealing in that it does not require specification of the missing data model, but one must model the effect of covariates on the response as well as the covariate distribution. It can yield quite efficient estimates of the marginal mean if both of these features are modeled correctly. Misspecification of the covariate distribution is less of a concern when covariates are discrete since a nonparametric estimate of this covariate distribution is available. It can become a serious concern, however, when one or more covariates are continuous. Discretizing continuous covariates is one approach for dealing with this problem, but one must then decide how coarsely to discretize the continuous covariates. Alternatively, non-parametric local likelihood methods [9] can be used for density estimation, but this may only be feasible if the number of continuous covariates is not too large.

The inverse probability weighted estimating equation approach does not require specification of the covariate distribution, and it gives a consistent estimator of the mean if the missing data model is correctly specified. In practice it can be easier to model the missing data process than the covariate distribution and model checks are possible based on binary regression modeling techniques. We found, however, that the IPWEE estimator may feature a loss of efficiency compared to the ECM approach.

We have focused on data from a cross-sectional studies as opposed to longitudinal studies. It is worth extending this investigation to the case where responses are collected over time in a longitudinal setting. In this case, covariates may be fixed as in this investigation, or time varying. In the latter case, multivariate models would be required to address the dynamics of the covariate process which makes the ECM approach considerably more challenging; the IPWGEE approach has considerable appeal in this case.

Robins et al. [13] proposed augmented estimating equations based on corrected complete-case analyses. A nice feature of the augmented approach is its "double robustness", meaning that the estimator obtained from the augmented method is asymptotically unbiased if either the underlying missing data mechanism *or* the underlying regression function is correctly specified. Furthermore, the augmented estimator can achieve full efficiency if both the missing data mechanism and the regression function are correctly specified. In general, however, it is very difficult to specify the regression function correctly, especially when the dimension of the covariate vector is high—this is the so-called curse of dimensionality problem. The augmented estimator can also have much lower efficiency if the working regression model is not close to the true regression model. See Kang and Schafer [6] for a review of the double robust estimator.

Multiple imputation is an alternative popular approach for dealing with incomplete data [8, 16, 17], but we have not explored this here. Qin and Zhang [11] discuss the idea of empirical likelihood estimation which is employed to seek a constrained empirical likelihood estimation of mean response with the assumption that responses are missing at random. The empirical-likelihood-based estimators enjoy the double-robustness property as well and it is also possible that empirical-likelihood-based inference can produce asymptotically unbiased and efficient estimators even if the true regression function is known. This is an intriguing approach warranting further research.

References

1. R.E. Bellman, *Adaptive Control Processes* (Princeton University Press, Princeton, 1961)
2. R. Cameron, K.S. Brown, J.A. Best, C.L. Pelkman, C.L. Madill, S.R. Manske, M.E. Payne, Effectiveness of a social influences smoking prevention program as a function of provider type, training method, and social risk. Am. J. Public Health **89**, 1827–1831 (1999)
3. P.J. Diggle, P. Heagerty, K.Y. Liang, S.L. Zeger, *Analysis of Longitudinal Data*, 2nd edn. (Oxford University Press, London, 2002)
4. G.N. Hortobagyi, R.L. Theriault, A. Lipton, L. Porter, D. Blayney, C. Sinoff, H. Wheeler, J.F. Simeone, J.J. Seaman, R.D. Knight, M. Heffernan, K. Mellars, D.J. Reitsma, Long-term prevention of skeletal complications of metastatic breast cancer with Pamidronate. J. Clin. Oncol. **16**, 2038–2044 (1998)
5. D.G. Horvitz, D.J. Thompson, A generalization of sampling without replacement from a finite universe, J. Am. Stat. Assoc. **47** 663–685 (1952)
6. J.D.Y. Kang, J.L. Schafer, Demystifying double robustness: A comparison of alternative strategies for estimating a population mean from incomplete data. Stat. Sci. **22**, 523–539 (2007)
7. K.Y. Liang, S.L. Zeger, Longitudinal data analysis using generalized linear models. Biometrika **73**, 13–22 (1986)
8. R.J.A. Little, D.B. Rubin, *Statistical Analysis with Missing Data* (Wiley, 2nd edn. 2002)
9. C.R. Loader, Local likelihood density estimation. Ann. Stat. **24**, 1602–1618 (1996)
10. P. McCullagh, J.A. Nelder, *Generalized Linear Models* (Chapman and Hall, London, 1989)
11. J. Qin, B. Zhang, Empirical-likelihood-based inference in missing response problems and its application in observational studies. J. Roy. Stat. Soc. B **69**, 101–122 (2007)
12. J.M. Robins, A. Rotnitzky, L.P. Zhao, Estimation of regression coefficients when some regressor are not always observed. J. Am. Stat. Assoc. **89**, 846–866 (1994)
13. J.M. Robins, A. Rotnitzky, L.P. Zhao, Analysis of semiparametric regression models for repeated outcomes in the presence of missing data. J. Am. Stat. Assoc. **90**, 106–121 (1995)
14. A. Rotnitzky, J.M. Robins, D.O. Scharfstein, Semiparametric regression for repeated outcomes with nonignorable nonresponse. J. Am. Stat. Assoc. **93**, 1321–1339 (1998)
15. D.B. Rubin, Inference and Missing data. Biometrika **63**, 581–592 (1976)
16. D.B. Rubin, *Multiple Imputation for Nonresponse in Surveys* (Wiley, New York, 1987)
17. J.L. Schafer, *Analysis of Incomplete Multivariate Data* (Chapman and Hall, New York, 1997)

Cardiovascular Informatics: A Perspective on Promises and Challenges of IVUS Data Analysis

Ioannis A. Kakadiaris and E. Gerardo Mendizabal Ruiz

Abstract Intravascular ultrasound (IVUS) is a catheter-based medical imaging modality that is capable of providing cross-sectional images of the interior of blood vessels. A comprehensive analysis of the IVUS data allows collecting information about the morphology and structure of the vessel and the atherosclerotic plaque, if present. Atherosclerotic plaque formation is considered to be a part of an inflammatory process. Recent evidence has suggested that the presence and proliferation of vasa vasorum (VV) in the plaque is correlated with the increase of plaque inflammation and the processes which lead to its destabilization. Hence, the detection and measurement of VV in plaque has the potential to enable the development of an index of plaque vulnerability. In this paper, we review the research at the Computational Biomedicine Lab towards the development of a complete pipeline for the detection and quantification of extra-luminal blood detection from IVUS data which may be an indication of the existence of VV.

Mathematics Subject Classification (2010): Primary 68U99, Secondary 68U01

I.A. Kakadiaris (✉)
Computational Biomedicine Lab, Departments of Computer Science, Electrical and Computer Engineering, and Biomedical Engineering, University of Houston, Houston, TX 77204, USA
e-mail: ioannisk@uh.edu

E.G. Mendizabal Ruiz
Computational Biomedicine Lab, Department of Computer Science, University of Houston, Houston, TX 77204, USA
e-mail: gerardomendizabal@gmail.com

P.M. Pardalos et al. (eds.), *Optimization and Data Analysis in Biomedical Informatics*, 117
Fields Institute Communications 63, DOI 10.1007/978-1-4614-4133-5_6,
© Springer Science+Business Media New York 2012

1 Introduction

Complications attributed to cardiovascular disease (CVD) constitute a major cause of death worldwide. One of the primary causes of CVD is coronary artery disease (CAD), which is a narrowing of the small blood vessels that supply blood and oxygen to the heart. CAD is caused by a condition called atherosclerosis, which occurs due to the accumulation of plaque on the inner walls of the arteries. The progression of this condition may lead to inflammation of the coronary arteries and the consequent obstruction of blood flow to the heart. But more critically, the sudden rupture of a plaque (i.e., thrombotic-related complications) may lead to a stenotic condition in which the blood supply is entirely cutoff from a region of the heart, resulting in death. In this context, the field of cardiology has introduced the term "vulnerable plaque" in reference to the plaques with a high likelihood of rupture, thrombotic complications, and the consequent rapid progression to stenosis [26–29]. Vasa vasorum ("vessels of the vessels", VV) is a network of microvessels that penetrates and "feeds" the vessel wall [13]. Recent evidence has suggested that the presence and proliferation (i.e., increase in density) of VV in the plaque is correlated to an increase in plaque inflammation and the processes which lead to its destabilization [1, 4, 7, 10, 12, 21, 22]. Hence, it is believed that the detection and measurement of VV in plaque and the detection of leakage of blood within atherosclerotic plaques have the potential to enable the development of an index of plaque vulnerability [3, 15].

Intravascular ultrasound (IVUS) is a catheter-based medical imaging technique that is capable of providing cross-sectional images of the interior of blood vessels and is currently the gold-standard technique for assessing the morphology of blood vessels and atherosclerotic plaques in vivo [42]. An IVUS system consists of a catheter with a miniaturized ultrasound probe attached to its tip. The ultrasound probe transmits ultrasound pulses and receives an acoustic radio frequency (RF) echo signal (i.e., A-line) at a discrete set of angles. A B-mode IVUS image is obtained by computing the positive envelopes of each A-line (Fig. 1a). The B-mode signals are compressed, stacked along the angular direction, and mapped into an 8-bit gray scale to form an image known as the polar B-mode image (Fig. 1b). To provide a more familiar representation of the data (i.e., one that resembles the interior of a vessel), the polar B-mode image is geometrically transformed to obtain a disc-shaped image known as the Cartesian B-mode image (Fig. 1c). Similar to other ultrasound modalities, IVUS may be used in combination with contrast agents [47] delivered as microbubbles which are of a size similar to red blood cells (diameter: $1–10\,\mu m$). These microbubbles resonate in response to the pressure changes induced by the ultrasound wave and are highly echogenic when compared to normal body tissues. As a result, they appear bright in the B-mode ultrasound images, and can hence be used as tracers of blood flow [3, 11].

Since VV may be found in the atherosclerotic plaque and/or the wall of the vessel (i.e., extra-luminal regions), the problem of VV detection can be posed as the detection of extra-luminal blood perfusion. In this paper, we present our studies

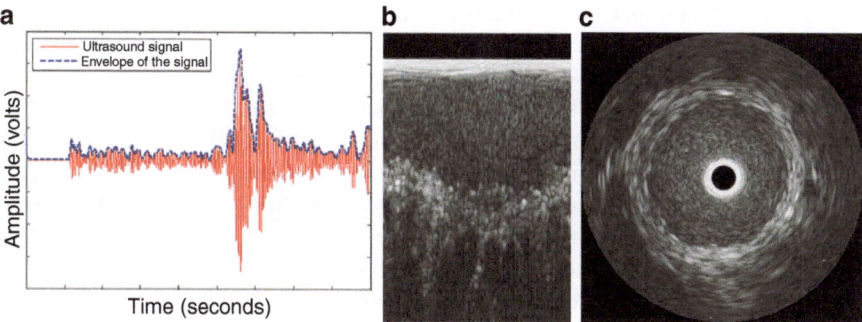

Fig. 1 Depiction of (**a**) A-line signal and its envelope, examples of the B-mode (**b**) polar, and (**c**) Cartesian IVUS images

towards the detection and quantification of extra-luminal blood perfusion, which can be categorized as: (1) methods for the detection of the lumen contour, and (2) methods for the detection of blood perfusion. The rest of this paper is organized as follows: In Sect. 2, we present a brief summary of methods that have been proposed for the analysis of IVUS data. In Sect. 3, we describe the methods for solving the extra-luminal blood detection problem. The results obtained with the proposed methods are presented in Sect. 4, and in Sect. 5, we present our conclusions.

2   Previous Work

IVUS Segmentation: Segmentation of IVUS data refers to the delineation of the lumen/intima and media/adventitia borders. This procedure is useful for studying atherosclerosis diseases, since it provides an assessment of the vascular wall, and also provides information on the nature of atherosclerotic lesions and information about the shape and size of the plaque. Automatic methods for IVUS segmentation are required as there are a large number of frames in an IVUS sequence, thereby making manual segmentation of a sequence infeasible (within a reasonable time). Some of the most recent approaches for automatic IVUS segmentation include a shape-driven method for lumen and media-adventitia segmentation introduced by Unal et al. [45] that uses Principal Component Analysis (PCA) to create a shape space from previously segmented frames. Segmentation is accomplished by the minimization of an energy function using nonparametric probability densities with global measurements. Taki et al. [44] proposed a method that involved preprocessing of the IVUS images, and the posterior deformation of geometric and parametric models using edge information. Downe et al. [6] introduced a method where PCA was first used for pre-processing. Active contour models were then used to provide an initial segmentation for a 3D graph search method. Multilevel discrete wavelet frame decomposition was used by Papadogiorgaki et al. [38] to generate texture information that was used along with the intensity information for contour initialization. Low pass filters and radial basis functions were then used to refine the contour. Similarly, Katouzian et al. [17] proposed a method where texture information was extracted using a discrete wavelet packet transform. The pixels of the IVUS image were then classified as lumen or non-lumen using k-means clustering. Finally, the contour was parameterized using a spline curve. Ciompi et al. [5] presented a method in which segmentation was tackled as a classification problem and solved using an error correcting output code technique. In this work, contextual information was exploited by means of conditional random fields computed from training data. The most common limitation of the segmentation methods discussed above is the lack of robustness with respect to noise, IVUS image variability, and the different artifacts that can appear in an IVUS image.

Tissue Characterization: Tissue characterization from IVUS data involves a definition of composition (e.g., fibrous, calcified, or lipid) of the atherosclerotic plaque based on the changes that occur to the sound waves as they interact with the different tissues. A common approach for characterization is to compute different texture features from the gray-level IVUS B-mode representation (e.g., co-occurrence matrix, laws features, Gabor filters). These features are used to train a classification model which is then used to predict the tissue classes on new data [2, 16, 41, 50]. The most successful approaches for the characterization of plaque are based on the analysis of the IVUS-RF signal data instead of the B-mode data. Nair et al. [30, 31] proposed a method known as "virtual histology" (IVUS-VH) that is based on the power spectral analysis (intercept, slope, mid-band fit, and minimum and maximum powers and their corresponding frequencies) of the IVUS-RF signals combined with classification trees. High accuracy (>85%) was reported for differentiating fibrous, fibrofatty, calcified, and necrotic regions. In addition, Rodriguez-Granillo et al. [40] and Nasu et al. [32], presented the results of *in-vivo* studies using the above method and reported a high correlation with the corresponding histology. Kawasaki et al. [18, 19] proposed a method for tissue classification using the integrated backscatter (IB), which is a parameter derived from the RF signal that is used to divide the tissue into five categories: thrombus, intimal hyperplasia or lipid core, fibrous tissue, mixed lesions and calcification. This method has demonstrated high sensitivity and specificity for characterizing calcification (100%, 99%), fibrosis (94%, 84%), and lipid pool (84%, 97%) [20]. O'Malley et al. [36] presented a study of the feasibility of blood characterization on IVUS data using features intended to quantify speckle and features based on frequency-domain measures of high-frequency signal using one-class support vector machines on the RF raw signal, the signal envelope and the log-compressed signal envelope. The feasibility of using wavelet analysis of the RF amplitude for plaque characterization [16, 41] and blood classification [17] was also studied. However, the majority of these methods are not suitable for blood detection since they focus on the characterization of the atherosclerotic plaque components. Also, the methods that have been proposed for blood detection are not capable of detecting small extra-luminal blood perfusion.

Perfusion Detection: O'Malley et al. [14, 34, 37, 46, 48, 49] proposed a protocol and an automatic algorithm (Analysis of Contrast Enhanced Sequences, ACES) for the quantification and visualization of VV in contrast-enhanced IVUS image sequences. That method relies on the detection of local echogenicity changes in stationary IVUS sequences caused by microbubble perfusion into the vessel wall. The proposed protocol consisted of acquiring images from a suspect plaque while a bolus injection of contrast agent was performed. The detection of extra-luminal blood was performed offline and involved two steps: (1) image stabilization [33,35] (i.e., image-based gating and registration), and (2) detection of enhancement, which was based on a comparison of the stabilized pre-contrast baseline images and the post-injection images. As a result, any change that occurred due to contrast enhancement would be reflected as a positive difference in the intensities. The enhancement was quantified and certain statistics were computed. The main limitation of this method is that it requires the alignment of frames which is very difficult to achieve even with the proposed stabilization methods. Goertz et al. [8, 9] proposed a solution for perfusion detection based on the detection of the harmonic and sub-harmonic response of the contrast microbubbles. The limitation of these methods is the requirement of a specially designed, non-commercial IVUS system.

3 Methods

Our proposed framework for the detection and quantification of extra-luminal blood perfusion consists of two steps: (1) detection of the luminal border, and (2) detection of extra-luminal perfusion. In the following subsections, we review the proposed methods for the above mentioned tasks.

3.1 Lumen Segmentation

Image-Based Segmentation: In this method, we employ the B-mode polar IVUS image representation for the segmentation of the lumen. This choice makes the computations much simpler due to the 1D appearance of the lumen contour. We define a function $f(\theta, c)$ as the curve that represents the change of interface between the lumen and the vessel wall. Since we know that the shape of the vessel's wall is essentially smooth, and that a polar B-mode IVUS image is periodic with respect to the horizontal axis, we parameterize the function that represents the lumen contour using Fourier series. The lumen segmentation problem consists of finding the optimum parameters c^* such that the curve $f(\theta, c^*)$ corresponds to the interface between the lumen and the vessel wall. This is accomplished by minimizing a cost function formulated using a Bayesian approach in which we incorporate *a priori* information about the regions of lumen and non-lumen based on the prediction of a support vector machine (SVM) classifier, trained with samples from the lumen and wall regions provided by the user [25].

RF-Based Segmentation: The use of B-mode images for IVUS data analysis poses a limitation due to the loss of information resulting from the B-mode conversion and the fact that the appearance of the B-mode images depends on the characteristics of the IVUS system which varies between systems, and on the visualization parameters (e.g., time gain compensation, compression, brightness, contrast) that are subjectively adjusted by the interventionist. To overcome this limitation, one has to work directly with the raw IVUS RF signal as it is not affected by the transformation or visualization parameters. Based on this observation, we developed a method for the segmentation of the lumen contour using the IVUS RF signal based on a physics-based model of the interaction of the sound waves with the tissues of the vessel [23].

When an incident sound wave interacts with an object, a fraction of its power will be reflected and a fraction will be absorbed by the object. When the wavelength of the incident wave is smaller in comparison with the size of the object, the reflection will occur in many directions (i.e., scattering). The power scattered by each scatterer object in the direction opposite to the direction of the incident wave is referred to as the *differential backscattering cross section* (DBC) [43]. If we consider that the wavelength of the IVUS impulse signal is large in comparison with the structures in the vessel, we can model the received IVUS RF signal, $\hat{S}_k(t)$, for each transducer's angular position (i.e., A-line) by representing the structures in the vessel as a finite set of point scatterers with an associated DBC coefficient. Our RF-based segmentation method consists of two steps: (1) a calibration step in which we estimate the parameters of the model using the RF signal of a manually segmented frame from the sequence to be segmented by employing an inverse problem approach, and (2) the detection of lumen contour by locating the change of interface for each A-line, by minimizing the cost function that employs the RF signal and the calibrated scattering model. Both the steps are based on the following assumptions: (1) there are only two types of tissue or layers within the vessel: lumen (blood) and wall; (2) the DBC coefficient of blood is different from the DBC coefficient of wall; (3) scatterers within the same layer will have the same DBC; (4) the attenuation coefficient is constant along the radial direction; and (5) the real IVUS signal can be approximated using a stochastic minimization process that employs random samples of the scatterers' positions. Since the lumen interface for each angle is recovered independently, it is very likely that the resulting curve is not smooth or periodic. Moreover, due to noise or artifacts, it is possible that our method obtains an incorrect result in one or more angles. Therefore, we introduce a post-processing step in which the lumen contour is constrained to a smooth periodical curve using Fourier series parameterization by applying a spectral smoothing method [39].

3.2   Perfusion Detection

Contrast Agent Detection: We investigated the feasibility of detecting the contrast
agent on IVUS sequences by characterizing the RF IVUS signal using two contrast
detection classifiers (CDC) based on one-class cost-sensitive learning [24]. In the
first contrast detection classifier (CDC_1), we build a model for the detection of
contrast agent using samples of the contrast agent present in the lumen during the
microbubble injection. In the second contrast detection classifier (CDC_2), we detect
the contrast agent as a change from baseline IVUS (i.e., lumen, intima, media and
adventitia signals acquired from frames prior to the bolus injection). The primary
advantage of these methods is that, by using the RF IVUS data, we do not lose
information contained in the frequency of the signal. The second advantage is that,
by using one-class learning, we do not need to provide "background" samples for
building the classifiers. This is particularly important to this study because, although
samples for contrast agent in lumen can be acquired by manual annotations from an
expert, the background can consist of a wide variety of other imaged tissues. Thus,
obtaining samples for the other tissues may be difficult and labor-intensive.

The features that characterize contrast agent and the baseline IVUS are defined
for a 3-D window of size $r \times \theta \times t$. These features are computed by stacking
consecutive frames over time, and obtaining a 3D IVUS signal volume $S(R, \Theta, T)$,
where R indicates the radial distance from the transducer, Θ is the angle with respect
to an arbitrary origin, and T is the time elapsed since the start of the recording
(i.e., frame number). We study the feasibility of characterizing the contrast agent's
signal using two types of features: features based on frequency-domain spectral
characterization (as proposed by O'Malley et al. [36]) and features based on 2-level
2D discrete wavelet decomposition.

Blood Detection: We assumed that that the signal of a partition corresponding to
certain tissue can be characterized by the DBC coefficient that generates that signal.
We employ the scattering model for computing the DBC corresponding to a partition
of the RF-signal of an A-line. Our objective is to find the DBC value that minimizes
the difference between the root mean square power (RMS) of the signal of a given
partition and the RMS power of the signal generated by our model. We divide the
real and modeled signal of each angle θ into N_P non-overlapping partitions of the
same size $\triangle P$. The initial and final times (α_p and β_p, respectively) for each partition
$P_{\theta,p}$ are computed such that $\triangle P = (\beta_p - \alpha_p) \forall p$, and the RMS of each partition of
the real and modeled signals ($R_{\theta,p}$ and $\hat{R}_{\theta,p}$, respectively) are computed. In order to
find the DBC value that generates the signal in each partition we find the value $\tau_{\theta,p}$
such that the quadratic error E between the RMS power of the real and modeled
signals for the partition $P_{\theta,p}$ is minimal. The RMS power of the modeled signal
depends on the spatial position arrangement of the scatterers. Since these positions
are unknown, we employ the Monte-Carlo approach on which we perform several
computations based on several random scatterer's positions. The problem of finding
the DBC for each partition is formulated as a system of linear equations and solved
very efficiently. Since we consider that there may exist a certain degree of overlap

between the ultrasound beams of consecutive angles due to the angular divergence of the beam, we introduce a regularization term that embodies our assumptions about the variation of the DBC value of each partition and its neighbors.

4   Results

Image-Based Segmentation: The first 50 frames from nine sequences of 20 MHz data (i.e., 450 frames in total) and nine sequences of 40 MHz data (i.e., 405 frames in total) were used for comparing the results of the automatic segmentation method with the manual segmentation from two expert observers. In order to evaluate the performance of the method, we computed the Dice similarity coefficient (degree of overlap between segmentation) along with linear regression and Bland-Altman analysis (comparison of lumen areas). The results indicated a high Dice similarity coefficient for the 20 MHz and 40 MHz datasets (0.95 and 0.93, respectively). The linear regression plots exhibited a high correlation between the measured areas obtained by the automatic and the manual segmentations. In addition, Bland-Altman analysis of the data indicated that the performances of the automatic method and the human observers are comparable. Figure 2 depicts examples of the segmentation results obtained with the proposed method.

RF-Based Segmentation: We evaluated the performance of the proposed method using the RF data from 490 frames corresponding to fourteen 40 MHz pullback IVUS sequences obtained from rabbit aortas and various coronary arteries of swine, and compared the results with those obtained through the the manual segmentation by expert observers. The average Dice similarity was 0.96 while the mean bias and the linear regression also showed that the performance of the automatic method and the human observers is comparable. Figure 3 depicts examples of the segmentation results obtained with the proposed method.

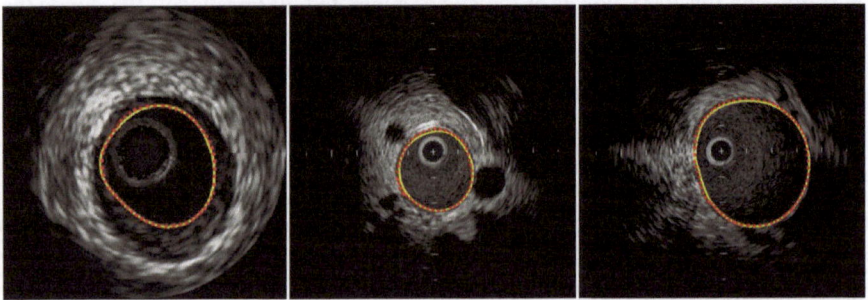

Fig. 2 Examples of automatic image-based segmentation results. The *solid* and *dotted lines* correspond to the automatic and manual segmentation, respectively

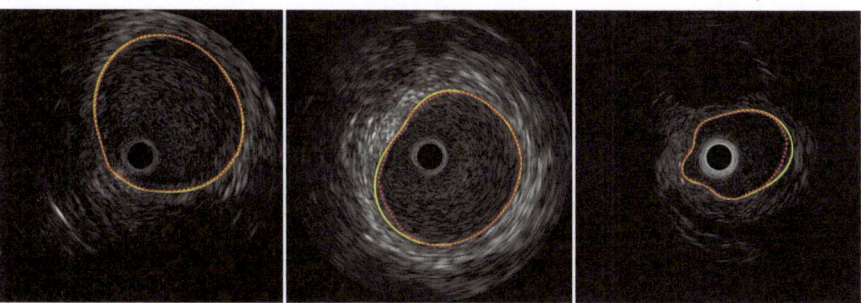

Fig. 3 Examples of automatic RF-based automatic segmentation results. The *solid* and *dotted lines* correspond to the automatic and manual segmentation, respectively

Contrast Agent Detection: Samples from two contrast-enhanced IVUS sequences obtained from swine were used to evaluate the feasibility of the proposed method. The best performance, for both CDCs and the two types of features, was obtained when using a window of size $r = 255$, $\theta = 7$, and $t = 13$. For the frequency-domain features, the best average performance for contrast detection (CD) and blood rejection (BR) with CDC_1 is CD = 96.61% and BR = 95.67%. With CDC_2 BD = 96.79% and CR = 94.24%. The best performance for wavelet-based features with CDC_1 is CD = 96.79% and BR = 94.13%. With CDC_2 BD = 98.51% and CR = 96.94%. Figure 4 depicts examples of the classification results obtained with the proposed method.

Blood Detection Results: Experiments were performed using real IVUS RF data from six 40 MHz pullback sequences corresponding to different arteries from rabbits and swines. For each sequence we compared the recovered DBC values for blood and non-blood samples acquired from manual annotations provided by an expert. The recovered DBC values for blood and non-blood were very similar for sequences acquired using the same IVUS system and from the same species. Additionally, as a preliminary blood detection experiment, we used our method to recover the DBC values from the IVUS RF data of a frame corresponding to a 40 MHz IVUS pullback from swine, for which histological information is available (Fig. 5a). We normalized the resulting DBC values for each pixel of the image and depicted the frame using a color palette (Fig. 5c). The regions of the resulting image that correspond to vascularization were manually annotated according to the criterion that a vessel should contain a region of DBC values corresponding to blood surrounded by DBC values corresponding to non-blood. These results are very encouraging as they provide preliminary evidence that our method could be used for computation of a feature that leads to automatic blood detection.

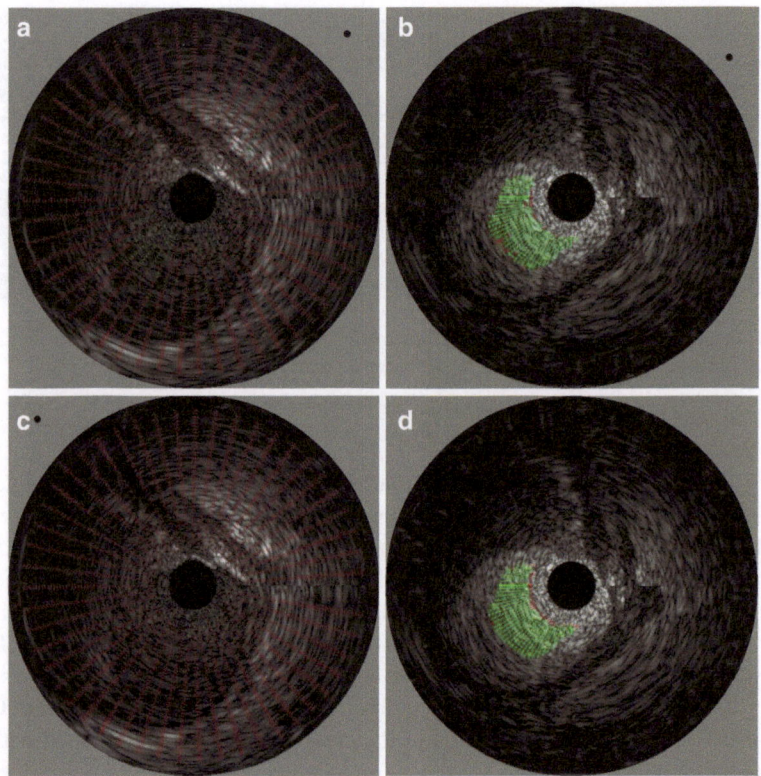

Fig. 4 Examples of classification results for CDC_1 using the frequency-domain-based (**a,b**) and wavelet-based features (**c,d**) in an IVUS frame before injection (**a,c**) and during the injection (**b,d**). The *green color* indicates the pixels classified as contrast agent and the *red color* indicates the pixels classified as non-contrast agent

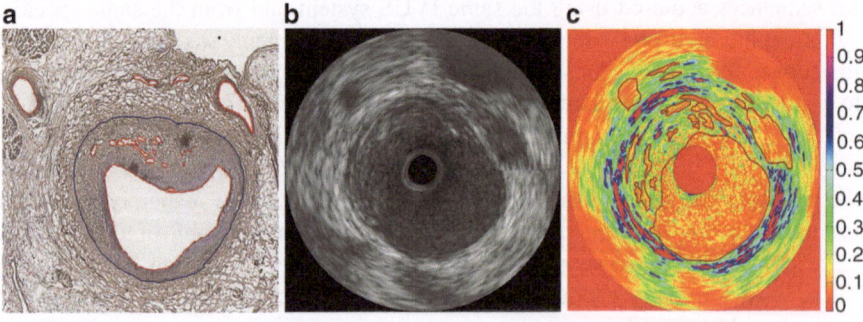

Fig. 5 Annotation of vasculature on (**a**) histological sample, (**b**) its corresponding B-mode Cartesian image, and (**c**) recovered DBC values using a color palette with annotation of blood regions

5 Conclusion

We have reviewed the proposed methods towards the development of a complete framework for the automatic detection of extra-luminal blood. The initial results of this study are very encouraging and we believe that further research in this direction will lead to the development of a fast and reliable method for VV detection and quantification.

Acknowledgements This work was supported in part by the Eckhard Pfeiffer Endowed Fund and by NSF grants IIS-0431144, CNS-0521527 and DMS-0915242. The second author was supported in part by CONACYT. Any opinions, findings, conclusions or recommendations expressed in this material are the authors' and may not reflect the views of the sponsors.

References

1. K.R. Balakrishnan, S. Kuruvilla, Role of inflammation in atherosclerosis: Immunohistochemical and electron microscopic images of a coronary endarterectomy specimen. Circulation **113**, e41–e43 (2006)
2. K. Caballero, J. Barajas, O. Pujol, J. Mauri, P. Radeva. In-Vivo IVUS Tissue Classification a Comparison Between Normalized Image Reconstruction and RF Signals Analysis. Proceedings of 11th Iberoamerican Congress on Pattern Recognition, Cancun, Mexico, pp. 137–146 (2006)
3. S. Carlier, I. Kakadiaris, N. Dib, M. Vavuranakis, C. Stephanadis, S. O'Malley, C. Hartley, R. Metcalfe, R. Mehran, E. Falk, K. Gul, M. Naghavi, Vasa vasorum imaging: A new window to the clinical detection of vulnerable atherosclerotic plaques. Curr. Atherosclerosis Rep. **7**(2), 164–169 (2005)
4. F. Chen, P. Eriksson, T. Kimura, I. Herzfeld, G. Valen, Apoptosis and angiogenesis are induced in the unstable coronary atherosclerotic plaque. Coron. Artery Dis. **16**(3), 191–197 (2005)
5. F. Ciompi, O. Pujol, E. Fernandez-Nofrerias, J. Mauri, P. Radeva, ECOC Random Fields for Lumen Segmentation in Radial Artery IVUS Sequences. Proceedings of 12th International Conference on Medical Image Computing and Computer Assisted Intervention, London, UK, pp. 869–876, 20–24 September 2009
6. R. Downe, A. Wahle, T. Kovarnik, H. Skalicka, J. Lopez, J. Horak, M. Sonka. Segmentation of Intravascular Ultrasound Images Using Graph Search and a Novel Cost Function. Proceedings of 2nd MICCAI Workshop on Computer Vision for Intravascular and Intracardiac Imaging, New York, NY, pp. 71–79, 10 September 2008
7. M. Fleiner, M. Kummer, M. Mirlacher, G. Sauter, G. Cathomas, R. Krapf, B.C. Biedermann, Arterial neovascularization and inflammation in vulnerable patients. Circulation **110**(18), 2843–2850 (2004)
8. D.E. Goertz, M.E. Frijlink, N. de Jong, A.F. van der Steen, Nonlinear intravascular ultrasound contrast imaging. Ultrasound Med. Biol. **32**(4), 491–502 (2006)
9. D.E. Goertz, M.E. Frijlink, D. Tempel, V. Bhagwandas, A. Gisolf, R. Krams, N. de Jong, A.F.W. van der Steen, Subharmonic contrast intravascular ultrasound for vasa vasorum imaging. Ultrasound Med. Biol. **33**(12), 1859–1872 (2007)
10. M. Gossl, N. Malyar, M. Rosol, P. Beighley, E. Ritman, Impact of coronary vasa vasorum functional structure on coronary vessel wall perfusion distribution. Am. J. Physiol. Heart Circ. Physiol. **285**(5), H2019–H2026 (2003)

11. J. Granada, S. Feinstein, Imaging of the vasa vasorum. Nat. Rev. Cardiol. 5(2s), S18–S25 (2008)
12. M. Hayden, S. Tyagi, Vasa vasorum in plaque angiogenesis, metabolic syndrome, type 2 diabetes mellitus, and atheroscleropathy: A malignant transformation. Cardiovasc. Diabetol. 3(1), 1–16, 2004
13. D.D. Heistad, M.L. Marcus, Role of vasa vasorum in nourishment of the aorta. Blood Ves. 16, 225–238 (1979)
14. I. Kakadiaris, S. O'Malley, M. Vavuranakis, S. Carlier, R. Metcalfe, C. Hartley, E. Falk, M. Naghavi, Signal processing approaches to risk assessment in coronary artery disease. IEEE Signal Process. Mag. 23(6), 59–62 (2006)
15. I. Kakadiaris, U. Kurkure, E. Mendizabal-Ruiz, M. Naghavi, Towards Cardiovascular Risk Stratification Using Imaging Data. Proceedings of 31st International Conference of the IEEE Engineering in Medicine and Biology Society, Minneapolis, MN, pp. 1918–1921, 2–6 September 2009
16. A. Katouzian, B. Baseri, E.E. Konofagou, A.F. Laine, An Alternative Approach to Spectrum-Based Atherosclerotic Plaque Characterization Techniques Using Intravascular Ultrasound (IVUS) Backscattered Signals. Proceedings of 2nd MICCAI Workshop on Computer Vision for Intravascular and Intracardiac Imaging, New York, NY, 2008
17. A. Katouzian, B. Baseri, E. Konofagou, A. Laine, Automatic Detection of Blood Versus Non-Blood Regions on Intravascular Ultrasound (IVUS) Images Using Wavelet Packet Signatures. Proceedings of SPIE Medical Imaging 2008: Ultrasonic Imaging and Signal Processing, San Diego, CA, 16–21 February 2008
18. M. Kawasaki, H. Takatsu, T. Noda, Y. Ito, A. Kunishima, M. Arai, K. Nishigaki, G. Takemura, N. Morita, S. Minatoguchi, H. Fujiwara, Noninvasive quantitative tissue characterization and two-dimensional color-coded map of human atherosclerotic lesions using ultrasound integrated backscatter: Comparison between histology and integrated backscatter images. J. Am. Coll. Cardiol. 38(2), 486–492 (2001)
19. M. Kawasaki, H. Takatsu, T. Noda, K. Sano, Y. Ito, K. Hayakawa, K. Tsuchiya, M. Arai, K. Nishigaki, G. Takemura, S. Minatoguchi, T. Fujiwara, H. Fujiwara, In vivo quantitative tissue characterization of human coronary arterial plaques by use of integrated backscatter intravascular ultrasound and comparison with angioscopic findings. Circulation 105, 2487–2492 (2002)
20. M. Kawasaki, B. Bouma, J. Bressner, S. Houser, S. Nadkarni, B. MacNeill, I. Jang, H. Fujiwara, G. Tearney, Diagnostic accuracy of optical coherence tomography and integrated backscatter intravascular ultrasound images for tissue characterization of human coronary plaques. J. Am. Coll. Cardiol. 48(1), 81–88 (2006)
21. F. Kolodgie, H. Gold, A. Burke, D. Fowler, H. Kruth, D. Weber, A. Farb, L. Guerrero, M. Hayase, R. Kutys, J. Narula, A. Finn, R. Virmani, Intraplaque hemorrhage and progression of coronary atheroma. New Engl. J. Med. 349(24), 2316–2325 (2003)
22. A.C. Langheinrich, A. Michniewicz, D.G. Sedding, G. Walker, P.E. Beighley, W.S. Rau, R.M. Bohle, E.L. Ritman, Correlation of vasa vasorum neovascularization and plaque progression in aortas of apolipoprotein E(-/-)/low-density lipoprotein(-/-) double knockout mice. Arterioscler. Thromb. Vasc. Biol. 26(2), 347–352 (2006)
23. E. Mendizabal-Ruiz, G. Biros, I.A. Kakadiaris, An Inverse Scattering Algorithm for the Segmentation of the Luminal Border on Intravascular Ultrasound Data. Proceedings of 12th International Conference on Medical Image Computing and Computer Assisted Intervention, London, UK, pp. 885–892, 20–24 September 2009
24. E. Mendizabal-Ruiz, I. Kakadiaris, One-Class Acoustic Characterization Applied to Contrast Agent Detection in IVUS. Proceedings of International Workshop on Computer Vision for Intravascular and Intracardiac Imaging, New York, NY, 10 September 2008
25. E. Mendizabal-Ruiz, M. Rivera, I. Kakadiaris, A Probabilistic Segmentation Method for the Identification of Luminal Borders in Intravascular Ultrasound Images. Proceedings of IEEE Computer Society Conference on Computer Vision and Pattern Recognition, Anchorage, AK, pp. 1–8, 24–26 June 2008

26. A.K. Mitra, A.S. Dhume, D.K. Agrawal, "Vulnerable plaques" – ticking of the time bomb. Can. J. Physiol. Pharmacol. **82**(10), 860–871 (2004)
27. M. Naghavi, P. Libby, E. Falk, S. Casscells, S. Litovsky, J. Rumberger, J. Badimon, C. Stefanadis, P. Moreno, G. Pasterkamp, Z. Fayad, P. Stone, S. Waxman, P. Raggi, M. Madjid, A. Zarrabi, A. Burke, C. Yuan, P. Fitzgerald, D. Siscovick, C. de Korte, M. Aikawa, K. Airaksinen, G. Assmann, C. Becker, J. Chesebro, A. Farb, Z. Galis, C. Jackson, I. Jang, W. Koenig, R. Lodder, K. March, J. Demirovic, M. Navab, S. Priori, M. Rekhter, R. Bahr, S. Grundy, R. Mehran, A. Colombo, E. Boerwinkle, C. Ballantyne, J. Insull, W., R. Schwartz, R. Vogel, P. Serruys, G. Hansson, D. Faxon, S. Kaul, H. Drexler, P. Greenland, J. Muller, R. Virmani, P. Ridker, D. Zipes, P. Shah, J. Willerson, From vulnerable plaque to vulnerable patient: A call for new definitions and risk assessment strategies: Part I. Circulation **108**(14), 1664–1672 (2003)
28. M. Naghavi, P. Libby, E. Falk, S. Casscells, S. Litovsky, J. Rumberger, J. Badimon, C. Stefanadis, P. Moreno, G. Pasterkamp, Z. Fayad, P. Stone, S. Waxman, P. Raggi, M. Madjid, A. Zarrabi, A. Burke, C. Yuan, P. Fitzgerald, D. Siscovick, C. de Korte, M. Aikawa, K. Airaksinen, G. Assmann, C. Becker, J. Chesebro, A. Farb, Z. Galis, C. Jackson, I. Jang, W. Koenig, R. Lodder, K. March, J. Demirovic, M. Navab, S. Priori, M. Rekhter, R. Bahr, S. Grundy, R. Mehran, A. Colombo, E. Boerwinkle, C. Ballantyne, W. Insull, R. Schwartz, R. Vogel, P. Serruys, G. Hansson, D. Faxon, S. Kaul, H. Drexler, P. Greenland, J. Muller, R. Virmani, P. Ridker, D. Zipes, P. Shah, J. Willerson, From vulnerable plaque to vulnerable patient: A call for new definitions and risk assessment strategies: Part II. Circulation **108**(15), 1772–1778 (2003)
29. M. Naghavi, E. Falk, H. Hecht, M. Jamieson, S. Kaul, D. Berman, Z. Fayad, M. Budoff, J. Rumberger, T. Naqvi, L. Shaw, O. Faergeman, J. Cohn, R. Bahr, W. Koenig, J. Demirovic, D. Arking, V. Herrera, J. Badimon, J. Goldstein, Y. Rudy, J. Airaksinen, R. Schwartz, W. Riley, R. Mendes, P. Douglas, P. Shah, From vulnerable plaque to vulnerable patient-Part III: Executive summary of the Screening for Heart Attack Prevention and Education (SHAPE) Task Force report. Am. J. Cardiol. **98**(2A), 2H–15H (2006); SHAPE Task Force Consensus Development Conference Journal Article Review United States
30. A. Nair, B. Kuban, E. Tuzcu, P. Schoenhagen, S. Nissen, D. Vince, Coronary plaque classification with intravascular ultrasound radiofrequency data analysis. Circulation **106**(17), 2200–2206 (2002)
31. A. Nair, M. Margolis, B. Kuban, D. Vince, Automated coronary plaque characterisation with intravascular ultrasound backscatter: Ex vivo validation. Eurointervention **3**(1), 113–120 (2007)
32. K. Nasu, E. Tsuchikane, O. Katoh, D. G. Vince, R. Virmani, J. F. Surmely, A. Murata, Y. Takeda, T. Ito, M. Ehara, T. Matsubara, T. Terashima, T. Suzuki, Accuracy of in vivo coronary plaque morphology assessment: A validation study of in vivo virtual histology compared with in vitro histopathology. J. Am. Coll. Cardiol. **47**(12), 2405–2412 (2006)
33. S. O'Malley, M. Naghavi, I. Kakadiaris, Image-Based Frame Gating for Stationary-Catheter IVUS Sequences. Proceedings of International Workshop on Computer Vision for Intravascular and Intracardiac Imaging, Copenhagen, Denmark, pp. 14–21, 1–6 October 2006
34. S. O'Malley, M. Vavuranakis, R. Metcalfe, C. Hartley, M. Naghavi, I. Kakadiaris, Intravascular Ultrasound-Based Imaging Of Vasa Vasorum for the Detection of Vulnerable Atherosclerotic Plaque. Proceedings of Houston Society for Engineering in Medicine and Biology, Houston, TX, 2006
35. S. O'Malley, S. Carlier, M. Naghavi, I. Kakadiaris, Image-Based Frame Gating of IVUS Pullbacks: A Surrogate for ECG. Proceedings of IEEE International Conference on Acoustics, Speech, and Signal Processing, Honolulu, HI, pp. 433–436, 15–20 April 2007
36. S.M. O'Malley, M. Naghavi, I.A. Kakadiaris, One-Class Acoustic Characterization Applied to Blood Detection in IVUS. Proceedings of 10th International Conference on Medical Image Computing and Computer Assisted Intervention, Brisbane, Australia, pp. 202–209, 29 October–2 November 2007

37. S. O'Malley, J. Granada, S. Carlier, M. Naghavi, I. Kakadiaris, Image-based gating of intravascular ultrasound pullback sequences. IEEE Trans. Inform. Tech. Biomed. **12**(3), 299–306 (2008)
38. M. Papadogiorgaki, V. Mezaris, Y. Chatzizisis, G. Giannoglou, I. Kompatsiaris, Image analysis techniques for automated IVUS contour detection. Ultrasound Med. Biol. **34**(9), 1482–1498 (2008)
39. J.R. Rice, and J.S. White, Norms for smoothing and estimation, SIAM Review, 6, pp. 243–256 (1964)
40. G.A. Rodriguez-Granillo, E.P. McFadden, M. Valgimigli, C.A. van Mieghem, E. Regar, P.J. de Feyter, P.W. Serruys, Coronary plaque composition of nonculprit lesions, assessed by in vivo intracoronary ultrasound radio frequency data analysis, is related to clinical presentation. Am. Heart J. **151**(5), 1020–1024 (2006)
41. A. Roodaki, A. Taki, S.K. Setarehdan, N. Navab, *Modified Wavelet Transform Features for Characterizing Different Plaque Types in IVUS Images: A Feasibility Study*. Proceedings of 9th International Conference on Signal Processing, Beijing, China, pp. 789–792, 26–29 October 2008
42. E. Sanidas, M. Vavuranakis, T. Papaioannou, I. Kakadiaris, S. Carlier, G. Syros, G. Dangas, C. Stefanadis, Study of atheromatous plaque using intravascular ultrasound. Hellenic J. Cardiol. **49**(6), 415–421 (2008)
43. K.K. Shung, M.B. Smith, B. Tsui, *Principles of Medical Imaging*. (Academic, NY, 1992)
44. A. Taki, Z. Najafi, A. Roodaki, S. Setarehdan, R. Zoroofi, A. Konig, N. Navab, Automatic segmentation of calcified plaques and vessel borders in IVUS images. Int. J. Comp. Assist. Radiol. Surg. **3**(3–4), 347–354 (2008)
45. G. Unal, S. Bucher, S. Carlier, G. Slabaugh, T. Fang, K. Tanaka, Shape-driven segmentation of the arterial wall in intravascular ultrasound images. IEEE Trans. Inform. Tech. Biomed. **12**(3), 335–347 (2008)
46. M. Vavuranakis, I.A. Kakadiaris, T.G. Papaioannou, S.M. O'Malley, S. Carlier, M. Naghavi, C. Stefanadis, Contrast-enhanced intravascular ultrasound: combining morphology with activity-based assessment of plaque vulnerability. Expert Rev. of Cardiovasc. Ther. **5**(5), 917–925 (2007)
47. M. Vavuranakis, I. Kakadiaris, S. O'Malley, C. Stefanadis, S. Vaina, M. Drakopoulou, I. Mitropoulos, S. Carlier, M. Naghavi, Detection of luminal-intima border and coronary wall enhancement in intravascular ultrasound imaging after injection of microbubbles and simultaneous sonication with transthoracic echocardiography. Circulation **112**(1), E1–E2 (2005)
48. M. Vavuranakis, T. Papaioannou, I. Kakadiaris, S. O'Malley, M. Naghavi, K. Filis, E. Sanidas, A. Papalois, I. Stamatopoulos, C. Stefanadis, Detection of perivascular blood flow in vivo by contrast-enhanced intracoronary ultrasonography and image analysis: An animal study. Clin. Exp. Pharmacol. Physiol. **34**(12), 1319–1323 (2007)
49. M. Vavuranakis, I. Kakadiaris, S. O'Malley, T. Papaioannou, E. Sanidas, M. Naghavi, S. Carlier, D. Tousoulis, C. Stefanadis, A new method for assessment of plaque vulnerability based on vasa vasorum imaging, by using contrast-enhanced intravascular ultrasound and automated differential image analysis. Int. J. Cardiol. **130**(1), 23–29 (2008)
50. D. Vince, K. Dixon, R. Cothren, J. Cornhill, Comparison of texture analysis methods for the characterization of coronary plaques in intravascular ultrasound images. Comp. Med. Imag. Graph. **24**(4), 221–229 (2000)

An Introduction to the Analysis of Functional Magnetic Resonance Imaging Data

Gianluca Gazzola, Chun-An Chou, Myong K. Jeong, and W. Art Chaovalitwongse

This paper is dedicated to our advisors.

Abstract Functional magnetic resonance imaging (fMRI) is a brain imaging technology primarily used to investigate how cognitive processes affect neural activity. Due to its non-invasiveness and high spatial resolution, this technology has quickly become one of the most important research tools in cognitive neuroscience and has played a growing role in a number of clinical applications. The interpretation of the results of an fMRI experiment involves the analysis of massive amounts of noisy, complex, multivariate data, resolved both spatially and temporally. The extraction of information from this data is a difficult and articulated task, which relies on methodologies lying at the intersection between image processing, statistics, and machine learning. We here introduce the reader

G. Gazzola
Rutgers Center for Operations Research, Rutgers University, Piscataway, NJ 08854-8003, USA
e-mail: ggazzola@rci.rutgers.edu

C.-A. Chou
Department of Industrial and Systems Engineering, University of Washington,
Seattle, WA 98115, USA
e-mail: joechou@uw.edu

M.K. Jeong
Rutgers Center for Operations Research and Department of Industrial and Systems
Engineering, Rutgers University, Piscataway, NJ 08854, USA
e-mail: mjeong@rci.rutgers.edu

W.A. Chaovalitwongse (✉)
Departments of Industrial and Systems Engineering and Radiology,
University of Washington, Seattle, WA 98195, USA
e-mail: artchao@uw.edu

P.M. Pardalos et al. (eds.), *Optimization and Data Analysis in Biomedical Informatics*,
Fields Institute Communications 63, DOI 10.1007/978-1-4614-4133-5_7,
© Springer Science+Business Media New York 2012

to the rich and diverse literature in the fascinating field of fMRI data analysis, providing an overview of its main challenges and of the most common approaches to overcome them.

Mathematics Subject Classification (2010): Primary 54C40, 14E20, Secondary 46E25, 20C20

1 Introduction

Magnetic resonance imaging (MRI) is a biomedical imaging technology that employs a combination of radio waves and strong magnetic fields to acquire detailed images of a subject's bodily structure via non-invasive scans. Functional magnetic resonance imaging (fMRI) emerged in the early 1990s as an evolution of conventional MRI specifically focused to mapping neural activity over time. fMRI uses the same hardware equipment as MRI, but it is unique in that it acquires sequences of individual MRI images to detect specific physiological changes in the scanned tissues. Local neural activity has been shown to be closely connected to local fluctuations in blood flow and blood oxygenation level. Due to a phenomenon known as Blood Oxygenation Level Dependent (BOLD) effect, when a neuron becomes active, its metabolic rate increases and so does its demand of oxygen, causing a shift in the relative concentration of oxygenated and deoxygenated hemoglobin in the blood flowing to the proximate vessels [1]. This quantity, which is referred to as BOLD signal or hemodynamic response, can be measured by an MRI scanner, since the magnetic susceptibility of blood is a function of its oxygenation level, and is therefore used as a proxy of neural activity [2].

fMRI has been largely used in cognitive neuroscience and psychology to study the neural bases of cognitive processes, by investigating how location and patterns of brain activity are affected by such conditions as the exposure to a sensory stimulus [3], the performance of a task [4], the control of emotions [5], the development of behaviors [6], or the making of a decision [7]. Following a somewhat complementary approach, neuroscientists have also fruitfully employed fMRI to learn from maps of spontaneous neural activations how different brain regions are functionally connected to each other [8]. Over the past decade, fMRI has played a growing role in clinical neuroimaging as a tool for pre-symptomatic diagnosis [9] and functional characterization [10] of neurological diseases. More recently, applications of fMRI in neurosurgical planning [11], neurorehabilitation [12], and drug discovery [13] and development [14] have also emerged.

The interpretation of the results of an fMRI experiment involves the analysis of massive amounts of noisy, complex, multivariate data, resolved both spatially and temporally. The extraction of information from this data is a difficult and articulated task, which relies on methodologies lying at the intersection between image processing, statistics, and machine learning. Our goal with this work is to introduce the reader to the rich and diverse literature in the fascinating field of fMRI data analysis, providing an overview of its main challenges and of the most common approaches to overcome them.

This chapter is structured as follows: Sections 2 and 3 describe how fMRI data is recorded, what information it contains, and the experimental designs according to which it can be acquired. Section 4 focuses on the numerous sources of noise in the data and on several of the possible preprocessing strategies for handling them. Sections 5 and 6 frame the exploration of fMRI data as a problem of statistical and machine-learning analysis, and introduce the most popular methodologies in the state of the art; An overview of the main software packages that implement such methodologies is provided by Sect. 7, which concludes this work.

2 Data Acquisition

During an fMRI experiment, a sequence of brain images are acquired while the subject performs one or more tasks laying inside the scanner. The measurements obtained by the scanner are collected in a 3-dimensional Fourier space, known as k-space. This data is then mapped into image space, possibly after undergoing some preprocessing. Since most types of analyses pertinent to the present chapter are carried out in image space, we refer the interested reader to [15] for a technical discussion on k-space.

Commonly, images are acquired along three planes: axial, coronal, and sagittal. Axial images are perpendicular to the vertical axis of the body (top-down), coronal images are parallel to the front of the body (front-back), and sagittal images are parallel to the side of the body (left-right). The physical size of the image, measured in mm^2, defines the field of view (FOV); the FOV is partitioned into uniform cells by a grid structure called acquisition matrix, whose size is typically 64×64 or 128×128; each of the cells in the grid is called a voxel, that is a volumetric pixel. The x- and y-dimensions of a voxel simply correspond to the ratio of the x- and y-dimensions of the FOV and the number of cells along the corresponding axes of the acquisition matrix; the z-dimension is given by the thickness of the brain slice mapped by the image. Typical values of voxel size are in the order $3 \times 3 \times 4$ mm.

An fMRI image records the BOLD signal at every voxel of a brain slice at a given time point (Fig. 1); a volumetric snapshot of the activity of the entire brain can be obtained by stacking together all images (typically in the order of 30 of them, each located at a different z-coordinate) acquired at the same moment. Throughout the course of an fMRI experiment, the activity of each voxel is measured at uniformly spaced time points (Fig. 2); the time distance between two successive observations is known as "repetition time" (TR). Typically, 100–2,000 observations are collected, with a TR ranging from 500 to 4,000 ms. In conclusion, the output of an fMRI experiment is a volume of time series, each of which describing the dynamics of neural activity at one of $\sim 10^5$ different brain locations (e.g., 30 z-slices defined on a 64×64 acquisition matrix yield a total of 122,880 voxels). Although a significant

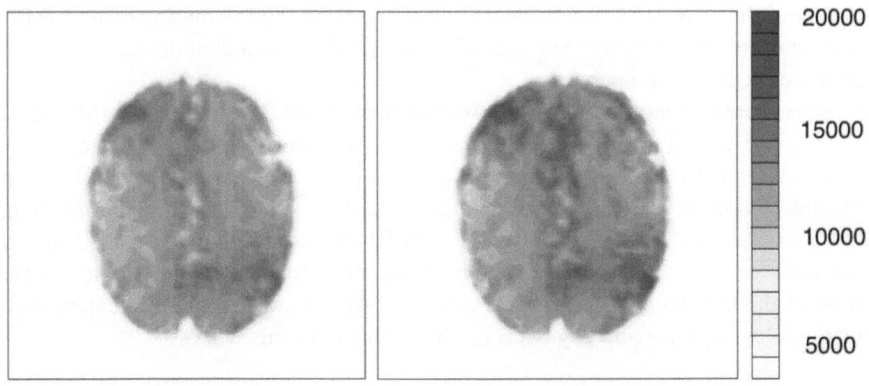

Fig. 1 BOLD signal of one axial slice, measured at two different time points during an experiment. Each image is defined by a 64×64 grid, composed by voxels with size $3 \times 3 \times 4.5$ mm; for every voxel, *darker colors* correspond to higher signal levels. Voxels corresponding to background noise outside the brain are shown in *white*. Note the overall increased activation in the *right-hand side* image

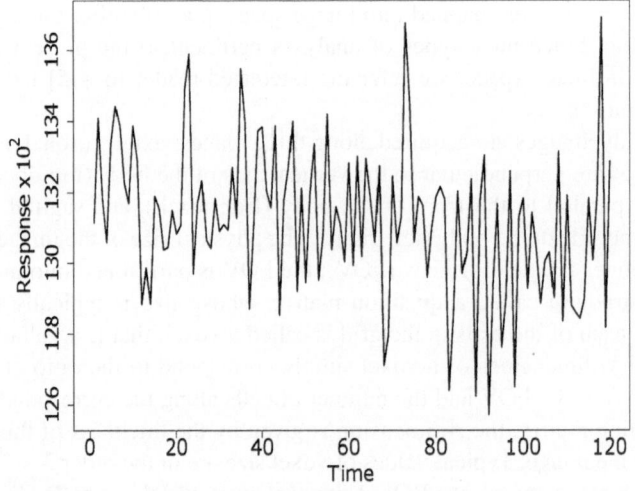

Fig. 2 BOLD signal of one individual voxel, measured at 120 equidistant time points

fraction of these voxels are "empty", that is they correspond to background noise outside the brain, and can be excluded (masked out) from the analysis, the overall amount of data to process is massive, especially if we consider that often the experiment is repeated multiple times for the same subject as well as for different subjects.

3   Experimental Designs

fMRI experiments are designed according to four main paradigms: the block design, the event-related design, the mixed design, and the behaviorally-driven design [16]. These designs differ in the scheme they adopt to expose the tested subject to cognitive tasks (Fig. 3), which in turn determines what specific research questions each design is most suited to answer.

The block design alternates periods or "blocks" of tasks with control blocks of rest, both with a fixed length. The subject performs the same type of task throughout every active block; for example, looking at a geometric shape on a display, or listening to a particular sound. A stimulus is generated for the entire duration of the task block, causing the hemodynamic response of the active voxels to accumulate and increase until reaching a plateau, declining back to baseline only when the following rest block starts. The response of voxels whose activity is not triggered by the stimulus remains unaffected. At the end of the experiment, the hemodynamic response at every voxel is averaged out across all blocks. Block designs have been shown to be robust to variability in the shape of the hemodynamic response [17] and to provide high statistical power for the detection of the subset of voxels activated by the task [18]. However, they are not suited for the precise estimation of transient features of the hemodynamic response.

In the event-related design, a sequence of different, possibly randomly chosen tasks are performed throughout the experiment [19]. The tasks events are usually shorter than in the block design. A given interstimulus interval (ISI)

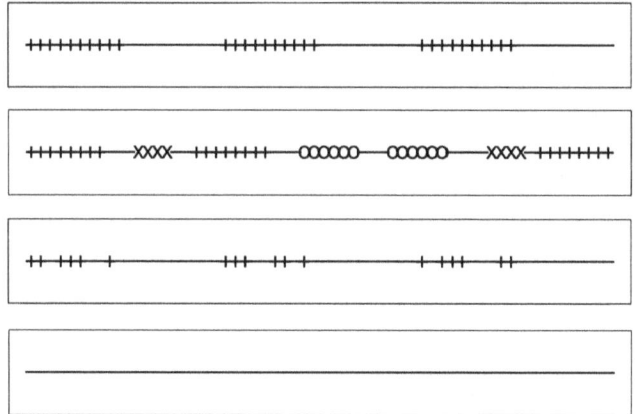

Fig. 3 From *top* to *bottom*: schematic representation of block, event-related, mixed, and behaviorally-driven design. Different *symbols* represent different tasks over time; a *straight line* represents rest. The block design (first graph on *top*) is defined as periodic sequence of task and rest blocks. The mixed-design (second graph) randomizes order and timing of different tasks. The mixed-design (third graph) alternates task and rest blocks like the the block design; within task blocks, however, tasks are performed multiple times with random timing. Finally, the behaviorally-driven design leaves the subject at rest for the entire duration of the experiment

separates the task periods; the ISI can be fixed or vary throughout the experiment. The unpredictability of the nature as well of the moment of presentation of the task results in a higher and more sustained attention level in the subject, thus reducing confounding effects caused by boredom or fatigue. Specularly to block designs, event-related designs are usually employed to accurately study how the shape of hemodynamic response varies depending on the specific task and active region being considered [20], but have lower statistical power for the detection of activations.

The mixed design is a hybrid of the previous two approaches, in which task blocks are alternated with control blocks and the ISIs within task blocks are randomized [21]. This type of design combines the advantages of both the block and the event-related design, in that it provides information regarding sustained as well as transient activity. For this reason, it is used both for activity detection and hemodynamic response estimation.

Differently from the previous three approaches, in the behaviorally-driven design the subject is not asked to engage in any task. Instead, the subject is simply instructed to lay inside the scanner in complete rest. During the experiment, spontaneous fluctuations of the BOLD response, also known as brain "resting-state" activity, are recorded. This information is then processed to infer how different regions of the brain functionally interact with each other, typically by the study of correlations and causal dependences between BOLD time series across the volume [22].

4 Noise Artifacts

4.1 Sources of Noise

One of the big hurdles in the analysis of fMRI data is determined by the presence of a significant amount of noise, coming from a number of different sources and exhibiting diverse statistical properties. Noise perturbations in fMRI data can be roughly categorized into four main groups: thermal noise, system noise, subject-related noise, and inter-subject noise. The former two are generated by physical fluctuations produced within the scanning process, whereas the latter are the effect of movement and physiological phenomena occurring while the subject is performing the experimental task.

Thermal noise is caused by variations in the thermal motion of electrons, both in the electronic components of the scanner and in the scanned tissues. This type of noise is task-independent, white in nature, and does not exhibit spatial structure across the volume; it can therefore be easily removed from the data by averaging out trial repeats.

System noise reflects different types of scanner instabilities and can be observed as a systematic, gradual drift in signal intensity over time. This drift can vary across voxels, both in magnitude and direction.

Although subjects are instructed to remain absolutely still during the scan, some degree of head movement is always observed, induced by the specific task being performed [23] or caused by actions like swallowing [24], blinking [25], or breath-holding [26]. These subject-related perturbations are spatially correlated and cause the signal from a given brain region to be contaminated with the signal from neighboring locations, therefore blurring the detected position of active and nonactive regions. Certain physiological functions can also induce motion-related noise [27], although generally smaller in scale, faster, and more predictable than the one produced by head movements. For example, respiration and cardiac cycles alter the position as well as the volume of the brain within the skull. This category of noise is pseudo-periodic in nature, it tends to have different effects on different brain regions, and is often found to be temporally correlated with neural activity.

Subject-related noise also includes neural activity patterns that are irrelevant for the purposes of the experiment. For example, it is known that eye movement results in activation of the frontal eye fields; the loudness of the scanner activates the auditory cortex and several brain regions related to attention [28]; mind wandering affects the temporal lobes and the frontal cortex [29].

Most experimental studies in fMRI are designed around a population of subjects. Although human brains share many anatomical regularities, a significant inter-subject variability is often observed in brain size, shape, and other features [30]. Due to these physiological fluctuations, the actual physical location represented by a given voxel might vary across images acquired on different subjects.

4.2 Noise Handling Strategies

Just a few percent points separate the signal intensity of an activated voxel from its baseline level. Due to their limited scale, changes in fMRI signal can be easily hidden by spurious perturbations. Not surprisingly, a vast amount of attention has been directed to the problem of noise handling in this field. There are two main strategies for reducing variability in fMRI data. The first and most natural one aims at preventing noise, by manipulating those among the confounding factors that are under the control of the experimenter. Part of the hardware-related noise, for example, can be eliminated by acting on the magnetic field strength and on the temperature of the scanning environment; head motion can be reduced, to a certain degree, by training the subject prior to the scan and by the use of various head restraints, such as bite bars, face masks, or cushions. Many sources of noise, however, are not controllable and even those that are cannot be eliminated entirely. Much of the work for variability handling is therefore carried out *a posteriori*, using a variety of approaches that aim at estimating and removing noise perturbations from the observed signal. The following paragraphs will focus on the main of these data-preprocessing procedures.

The acquisition of a 3-dimensional image is not carried out as a whole but rather as a superimposition of a sequence of individual 2-dimensional slices. This sequence can be ascending or descending, and can possibly take place in an interleaved fashion: for example, with all odd slices acquired first, followed by all even ones. A slight time delay occurs between the acquisition of every two consecutive slices, causing a non-negligible time misalignment to occur in the data. Slice time correction is a kind of data interpolation that is used to correct for this source of error [31]. Typically, the correction consists of realigning the data of all slices to the time point at which a given reference slice (for example, the first one) was recorded, by resampling the data at time points falling in between measurements. The values at the non-measured time points are estimated by means of data points measured in their proximity, using a variety of linear and nonlinear (for example, splines, sinc, etc.) interpolation methods.

The drift in the signal introduced by system noise is handled through different forms of detrending. The simplest approach is mean correction, which consists in readjusting all images corresponding to the same slice acquired during a study so that they all share the same average intensity. More sophisticated techniques explicitly estimate the drift by means of linear, polynomial, spline, or wavelet models of low-frequency noise [32].

Head motion is a major issue in data preprocessing, since even minimal movements can induce significant measurement artifacts, potentially of greater magnitude than the true signal itself. The typical approach to handling head-motion noise starts with the alignment of each image to a reference image, for example, the first one or the average of all those acquired. A common assumption is that head and brain move as a rigid body, that is their shape remains constant while their position and orientation in space changes. With this assumption, the motion estimation problem reduces to assessing the magnitude of rotations and translations along the three axes. Most estimation methods define each such transformation with three different parameters and aim at finding the values of those parameters that minimize the distance between the input image and the one chosen as reference. This is done by means of an array of iterative optimization techniques and distance measures, such as least squares, absolute difference [33], or mutual information [34]. Once the motion parameters have been optimized, they can be used to register the input image to the reference target; this correction step makes use of spatial interpolation methods (for example, linear, polynomial, sinc, etc.), since signal values need to be calculated at positions falling in-between measured data points.

Some of the proposed strategies for handling physiological noise involve monitoring the actual physiological motion of the subject simultaneously as the neural activity is being measured during the scan. The latter is then synchronized to the former by matching each measured point to the cardiac or respiratory cycles it belongs to. Finally, physiological fluctuations are estimated with a variety of approaches (for example, Fourier series) and removed from the data [27, 35].

Other techniques focus solely on the BOLD signal, by looking for the dominant frequency components within the typical respiratory and cardiac frequencies [36]. These methods do not rely on the external recording of physiological cycles, but their effectiveness requires the TR to be short, so that physiological noise can be sampled without aliasing.

Inter-subject fluctuations are corrected by means of a normalization procedure, in which images acquired on different subjects are registered to a common reference space. This procedure makes use of affine or non-linear transformations [37] that interpolate input images and warp them around a template brain (most commonly, the Talairach [38] or the Montreal Neurological Institute brain [39]).

Image spatial smoothing is a preprocessing technique employed to improve the signal-to-noise ratio in the data, independently of the specific factor inducing noise. Spatial smoothing is typically obtained convolving images with a Gaussian kernel, usually characterized by a parameter named "full width half maximum" (FWHM), which defines the kernel width at half its maximum height [40]. Common values of the FWHM are in the range of 4–10 mm; the specific value must be accurately selected in order to avoid, on the one hand, blurring out the activity of very small regions or artificially merging proximate active regions that are functionally separated from each other (which would occur if the FWHM is too large), and, on the other hand, insufficient noise compression or degradation of spatial resolution (which would take place if FWHM is too small).

5 Neural Activity and Basics of Statistical Analysis

Changes of the BOLD signal in response to a single stimulus generated by a specific task theoretically follow a temporal pattern known as hemodynamic response function (HRF). The standard canonical model for the HRF is a smooth curve, which remains constant for approximately 2 s after the application of the stimulus and then gradually increases until reaching a peak 5–8 s later. Assuming no further stimulus is applied, the signal takes about 15–20 s to decay back to baseline levels. During the last 10 s, a dip under baseline is observed [41]. Several studies have given indication of nontrivial deviations between empirically measured HRFs and the canonical model; for example, a dip below baseline is often detected prior to the initial rise of the signal [42]. For this reason, more complex and flexible models of the HRF have also been proposed, based, for example, on Gamma [43] or Gaussian [44] functions.

Once a theoretical form has been chosen for the HRF, the modeled BOLD signal at time t $b(t)$ is obtained by convolving the stimulus function (typically a "box-car" function that encodes the occurrence of a stimulus in time) with the HRF model, that is $b(t) = s(t) \otimes h(t)$, where $s(t)$ and $h(t)$ are the stimulus function and the theoretical HRF at time t, respectively (Fig. 4). This convolution is usually carried out in a linear fashion, assuming that the BOLD signal is simply the sum of successive responses.

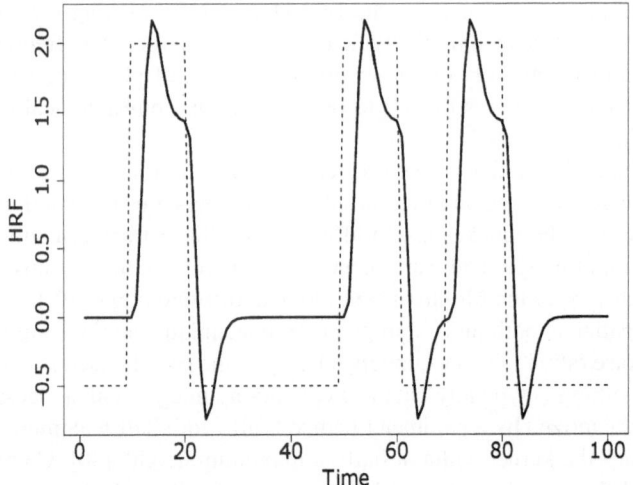

Fig. 4 Hemodynamic response function (*solid line*), modeled as a difference of two Gamma functions with parameters as in [43], convolved with a stimulus function (*dashed line*) with three 10-s stimuli, starting at times 10, 50, and 70

Traditionally, the statistical analysis of fMRI data orbits around the use of general linear models (GLM) to estimate the relationship between the observed BOLD signal of a given voxel and a "design matrix" of explanatory variables, each of which assumed to represent a different signal component [45]. These inputs include terms describing the time of occurrence of different stimuli, as presented by the chosen design of experiments, as well as the theoretical BOLD response and nuisance factors, such as signal drifts and periodic components. This problem is univariate, since it is defined on a voxel-by-voxel basis, and consists in computing an Ordinary-Least-Square optimal set of β parameters for the model

$$Y = \beta X + \epsilon, \tag{1}$$

where Y is the observed BOLD time series, X is the design matrix, and ϵ is an error term, usually assumed to be normally distributed with mean 0 and variance σ^2. The significance of the model parameters at every voxel is then assessed running a statistical test; the distributed outputs of all tests (for example, t-values) are then used to build a statistical parametric map (SPM) across the voxel volume. The SPM is finally manipulated, usually through a thresholding process based on Gaussian Random Field (GRF) theory, to extract the subset of voxels that show the highest systematical response to a given covariate (for example, voxels whose activation is most strongly triggered by a specific type of task).

Multivariate techniques are also commonly used as exploratory tools of fMRI data, with Principal Component Analysis (PCA) and Independent Component Analysis (ICA) [46] among the most popular. PCA expresses the data as a weighted sum of uncorrelated components, in order to identify spatial patterns that are responsible for the most variability of the BOLD time series across voxels. This is obtained via the well-known Singular Value Decomposition (SVD) of the t x n data matrix (with t being the number of time observations for each of n voxels), that is:

$$X = UDV^T, \tag{2}$$

where U and V^T are a t x t and an n x n unitary matrix, representing temporal and spatial components of the data, respectively, and D is a t x n diagonal matrix with non-negative elements on the diagonal representing the amount of variability captured by each component.

Similarly to PCA, ICA operates a decomposition of the data matrix X into a linear combination of variables, which however are required to be spatially statistically independent. The specific transformation is defined as:

$$X = MS, \tag{3}$$

where S is a m x n matrix containing m spatially independent source signals, and M is a t x m mixing matrix of weights. The same procedure applied to X^T allows to extract temporally independent components.

Advanced multivariate analysis techniques for fMRI pattern analysis will be covered by the following section, where our attention will be focused on methodological issues that are more exquisitely related to the science of machine learning.

6 Machine-Learning Analysis

Over the past decade, there has been a growing interest in the use of machine-learning methods as tools for multivariate analysis of fMRI data. How to characterize distributed patterns of neural activity generated by different stimuli, predict cognitive states from a set of image features, or partition the gray matter into areas with homogeneous temporal behavior, are just some of the myriad research questions that have been successfully addressed by means of a variety of machine-learning algorithms, both in a supervised and unsupervised fashion. In this section we will give an overview of the main problems and approaches described in the rich literature in this field.

6.1 Feature Extraction and Selection

The first step toward the application of machine learning tools for the extraction of patterns from an fMRI data set is the definition of a set of informative and robust features to be used as input. The simplest approach to this problem is to collect one data point for every 3-dimensional image acquired, and use the activation level of every voxel as feature. This technique has the advantage of providing many observations for training and testing the classifier, but its main drawback is that a significant amount of noise might be present in the observation, even after the preprocessing stage. For such reason, many authors prefer to use fewer but cleaner observations, extracted from spatial or temporal averages of voxel activity. For example, in [47] the authors divide the brain into a certain number of regions of interest (ROI) and define the average activation of the voxels within each ROI as one feature; in [48] features are obtained by averaging out the activity of individual voxels over each task block of the experiment.

More elaborate approaches aim at synthesizing features from the original signal, using different types of transformations. One common idea is to map each image into its components via SVD [49] or spatial/temporal ICA [50]. Several authors have also used basis projections of the activity time series based on wavelets [51], splines [52] and Fourier bases [53]. Others proposed the use of a GLM regressor to fit the activity time series of each voxel and use the pattern of the estimated β parameters and t-values across all voxels as features [54].

fMRI data is generally characterized by a number of features that largely exceeds the number of observations available. In order to remove uninformative and redundant inputs and reducing the risk of overfitting the data when training statistical models, a great deal of attention has been focused on feature selection procedures. In some cases, feature selection is carried out simply based on a priori criteria: the experimenter might for example want to restrict the analysis to a few specific ROIs, defined in terms of either functional or anatomical properties [55]. When the selection of features is to be carried out algorithmically, practitioners often make use of univariate scoring approaches, which rank voxel features by a specific criterion [48]. These criteria include: scoring voxels based on a t-test on the difference between their mean activity level during task blocks and a control baseline; training a classifier on one only voxel at a time and scoring every voxel based on the resulting classification accuracy; ranking voxels based on how consistently they react to different stimuli across cross-validation groups; using p-values returned by analysis of variance to select voxels whose mean activity varies most significantly across stimuli. Multivariate techniques are also commonly used. Some authors introduced a variation of the one-voxel-at-a-time scoring criterion described above named "searchlight", which ranks according to the accuracy of a classifier trained on a local neighborhood of every given voxel [56]. Others proposed dimensionality reduction techniques based on SVD [57] and ICA [58].

6.2 Classification

In the fMRI framework, classification can be described as the supervised modeling technique that aims at predicting the cognitive state of the brain from distributed activation patterns [59]. Activation patterns generated by different categories of stimuli can be distinctly discriminated by classifiers. This has been shown by numerous studies applied to experiments on motor, perceptual, and cognition-related tasks. For example, in [60] the authors describe an experiment consisting of button press blocks in which the subject was asked to alternately use the left or right index finger. A classifier was then tested to determine in real time which of the two tasks the subject was performing. In [61] a group of subjects was presented sequences of images showing different categories of objects. Classifiers were then applied to determine the type of object the subjects were observing, using small subsets of the collected data as input. Others [47] explored the possibility of classifying activation patterns evoked by the presentation of ambiguous versus unambiguous sentences. Earlier studies focused on the problem of single-subject classification, that is training and testing a model on data belonging to the same experimental subject. More recently, classification of fMRI data was successfully extended to a multi-subject framework [62].

One of the primary goals when analyzing fMRI data from a supervised perspective is the decoding of activation patterns to develop insight on how the brain works. Thus, it is desirable that the decision rules discriminating classes are simple enough to be interpretable by practitioners. For such reason, the classifiers most frequently used in the fMRI literature fall under the broad category of linear models, with linear Support Vector Machines (SVM), [63], Fisher's Linear Discriminant Analysis (LDA) [61], and Gaussian Naïve Bayes (GNB) [64].

Let X be an $n \times m$ input fMRI data matrix, X_i be one of its n observations (for example, neural activity at a given time point) defined on m features (for example, all voxels in the volume), and Y_i be the class label of such observation. In its soft-margin implementation, SVM estimates an optimal linear discriminant by solving the following mathematical program [65]:

$$\min_{w,b,\xi} \frac{1}{2} \| w \|^2 + \lambda F \left(\sum_{i=1}^{n} \xi_i \right) \tag{4}$$

$$\text{s.t. } Y_i(X_i w - b) \geq 1 - \xi_i, \quad \xi_i \geq 0, \quad i = 1, 2, \ldots, n,$$

where w and b are the vector of m weights and the bias term of the linear discriminant, respectively, λ is a non-negative regularization constant, ξ_i is a slack variable that penalizes the misclassification of observation X_i, and $F(\cdot)$ is a monotonic convex function, whose form is chosen so that (4) is a quadratic program.

The derivation of both LDA and GNB [46] is based on the Bayes' rule, which for two random variables X and Y holds that:

$$P(Y = k|X) = \frac{P(X|Y = k)P(Y = k)}{P(X)}. \tag{5}$$

With X and Y representing m-dimensional data points and class labels, respectively, LDA models each class k as an m-dimensional Gaussian with mean μ_k and covariance matrix Σ_k such that:

$$P(X|Y = k, \mu_k, \Sigma_k) = \frac{1}{(2\pi)^{m/2}|\Sigma_k|^{1/2}} \exp\left(-\frac{1}{2}(X - \mu_k)^T \Sigma_k^{-1}(X - \mu_k)\right), \tag{6}$$

where $\Sigma_k = \Sigma$ for all classes. This results in a class conditional probability

$$P(Y = k|X) \propto \exp\left(-\frac{1}{2}(X - \mu_k)^T \Sigma_k^{-1}(X - \mu_k)\right) P(Y = k), \qquad (7)$$

and a linear discriminant function

$$f_k(X) = X^T \Sigma_k^{-1} \mu_k - \frac{1}{2}\mu_k^T \Sigma_k^{-1} \mu_k + \ln P(Y = k). \qquad (8)$$

Class probability, mean, and covariance matrix are estimated as follows:

$$P(Y = k) = n_k/n,$$
$$\mu_k = \sum_{i:Y_i=k} X_i/n_k, \qquad (9)$$
$$\Sigma_k = \Sigma = \sum_{k=1}^{K} \sum_{i:Y_i=k} (X_i - \mu_k)(X_i - \mu_k)^T/(n - K),$$

where $n_k = |\{i : Y_i = k\}|$ and K is the total number of classes.

GNB is a variant of LDA, which additionally assumes that the conditional distribution of each feature is independent of the others, that is:

$$P(X|Y = k) = \prod_{j=1}^{m} P(X^j|Y = k), \qquad (10)$$

where X^j defines the j-th feature. If we further assume that the variance of each feature is the same across all classes (which is often the case in fMRI data), GNB can also be expressed as a linear discriminant [48].

Several authors showed that the linear approaches tend to yield higher classification accuracy than non-linear ones [54]. This phenomenon has been interpreted as an effect of overfitting due to the sparsity of the data, rather than of the intrinsic lack of real interactions between individual voxels [48].

Systematic comparisons of different classification methods have been performed in a number of studies. SVM appears to be the most promising modeling technique, shown to outperform benchmark methods in most studies. A commonly emphasized caveat is that non-linear kernels tend to undermine the robustness of the model to experimental noise [61,63]. In [47], the authors applied SVM, GNB, and k-nearest neighbors classifiers to three case studies, obtaining higher classification accuracy with both SVM and GNB. LDA has also been proven to be a valid and computationally appealing alternative to SVMs. This is particularly true in lower-dimensional problems, where the inversion of the estimated covariance matrix operated by LDA can be significantly less computer intensive than the cross-validation-based search for optimally tuning the parameters of SVMs. A recent study reported encouraging results for a classification method based on regularized logistic regression, which performed better than SVMs on both simulation experiments and real data [66].

6.3 Clustering

In the fMRI literature, the problem of clustering is often formulated as that of partitioning a volume of data into voxels sets with homogeneous temporal behavior, under the assumption that voxels with similar activation patterns are likely to share the same functional properties [67, 68]. The search for similarities can be applied to the raw time series, for example via the computation of the Pearson correlation coefficient between pairs of voxels [69]. Some studies, however, show that such approach tends to provide unstable results due to the intrinsic noisiness of the data; moreover, it does not account for similarities between time series that are not related to the task. These observations justify alternative approaches, in which comparisons between voxels are made on transformations of their respective time series rather than on the time series themselves. For example, in [70] the cross-correlation function of the activity signal with respect to the experimental protocol signal (e.g., the box-car time series of a block design) is used to drive voxel grouping; in [71], the authors emphasize the benefits of clustering on the signal autocorrelation function in situations where not all active voxels start responding to the stimulus at the same time.

Since usually only a small portion of the brain gets activated during an experiment, many authors recommend to screen the volume to eliminate completely unresponsive voxels from the input of the cluster algorithm, in order to increase the chances that the returned clusters are actually expression of different groups of activation [72]. Some authors also suggest to mask out all voxels that are not part of the gray matter [73].

The most popular clustering tools in fMRI data analysis are k-means, both in its standard [70] and fuzzy [73] variant, and agglomerative hierarchical clustering, the latter possibly in combination with sharpening methods [74].

Let X be a n x m input fMRI data matrix and X_i one of its n data points (that is, one voxel), defined on m features (for example, all time observations of neural activity). The k-means algorithm [75] begins with the initialization of k points $c_1, c_2, \ldots, c_k$, each of which defines the centroid of a different cluster. Subsequently, an assignment and an update phase are iterated until convergence. In the assignment phase, each point X_i is assigned to the cluster with the closest centroid with respect to a metric d, that is cluster C_j is computed as follows:

$$C_j = \left\{ X_i : d(X_i - c_j) \leq d(X_i - c_l) \right\}, \forall l \neq j. \tag{11}$$

In the update phase, the mean of the points in cluster C_j becomes the new centroid, that is:

$$c_j = \frac{1}{|C_j|} \sum_{i:X_i \in C_j} X_i \qquad (12)$$

The initial centroids are typically obtained either by randomly assigning points to k different clusters and then applying formula (12), or by randomly choosing k points from X. Convergence is reached when cluster assignments are stable.

In the update phase of the fuzzy variant of the k-means algorithm [76] the centroid c_j of cluster C_j is computed as the mean of all points $X_i \in X$, each of which weighted by the degree u_j of its belonging to such cluster, that is:

$$c_j = \frac{\sum_{i=1}^{n} \left(u_j(X_i)\right)^r X_i}{\sum_{i=1}^{n} \left(u_j(X_i)\right)^r}, \qquad (13)$$

$$u_j(X_i) = \frac{1}{\sum_{l=1}^{k} \left(\frac{d(X_i - c_j)}{d(X_i - c_l)}\right)^{2/(r-1)}}, \qquad (14)$$

where $r > 1$ is a real parameter that controls cluster fuzziness. Convergence is reached when $\max_{i,j} |u_j^{t+1}(X_i) - u_j^{t}(X_i)| < \epsilon$, with ϵ being a real parameter with value close to zero and t and $t+1$ being two successive iterations.

Agglomerative hierarchical clustering algorithms [46] build a tree-like hierarchy of clusters by means of a bottom-up approach, which starts by creating a set of n singleton clusters (each defined by a different point in X) that are then recursively merged together by pairs until one single cluster (including all n points) is obtained. The clusters merged at every recursion are the two with the least dissimilarity, as measured by a metric known as linkage. This metric can take different forms, the most commonly used ones of which are: the single or nearest-neighbor linkage, defined as

$$\min \left\{d(X_i, X_j) : X_i \in C_v, X_j \in C_w\right\}, \qquad (15)$$

the complete, or farthest-neighbor linkage, given by

$$\max \left\{d(X_i, X_j) : X_i \in C_v, X_j \in C_w\right\}, \qquad (16)$$

and the group-average linkage, expressed as

$$\frac{1}{|C_v||C_w|} \sum_{i:X_i \in C_v} \sum_{j:X_j \in C_w} d(X_i, X_j). \qquad (17)$$

In fMRI data analysis, the clustering metric is usually a negative function of the correlation between the (raw or transformed) signals of a pair of voxels [71]; Some authors also proposed an aggregate metric that depends both on correlation and geometrical proximity, considering that activity of local, functionally related brain regions are known to exhibit similar temporal behavior [77].

The problem of choosing the optimal number of clusters has been addressed in different ways. In [78], a two-phase technique is used, which first applies a hierarchical technique to determine the number of clusters and subsequently uses the outputted clusters as initial condition for a k-means algorithm. Other approaches use validation methods that aim at assessing the significance of the identified clusters via statistical testing [79] or at finding clustering patterns that generalize best across a set of subjects based on a cross-validation likelihood measure [72].

7 Computer Software for fMRI Data Analysis

In this conclusive section we provide an overview of some among the most popular software packages for the analysis of fMRI data, selected from those freely downloadable from the Internet. A comprehensive list of computational resources, both for fMRI and other neuroimaging applications, can be found at the Neuroimaging Informatics Tools and Resources Clearinghouse (www.nitrc.org) and the Internet Analysis Tools Registry (www.cma.mgh.harvard.edu/iatr/) websites.

SPM (www.fil.ion.ucl.ac.uk/spm/) is by far the most widely used package for univariate analysis of fMRI data. Written as a Matlab suite, it includes a variety of routines for data preprocessing (time-slice correction, realignment, spatial smoothing, etc.) and is designed both for single-subject and group studies. The analysis is carried out with a standard GLM approach, in which the parameters of a specified general linear model are estimated from the observations and the results are used to compute a statistical parametric map. The user can control many analytical details, such as the HRF form (for example, Canonical, Gamma-function-based, etc.) and the techniques for the estimation of model parameters (maximum likelihood, Bayesian). Another prominent package for univariate analysis is *AFNI* (afni.nimh.nih.gov/afni), which, besides several features shared with *SPM*, includes, among others, functions for mapping images to Talairach coordinates, display data as axial, coronal and sagittal slices and compute activation maps with correlation methods. *AFNI* is written in C and can be run both via a graphical user interface and as a batch process. The fMRI toolboxes of the *FSL* library (www.fmrib.ox.ac.uk/fsl/) are also a popular alternative to *SPM*: in particular, *FEAT* for data preprocessing and GLM analysis and *FLOBS* for HRF generation and Bayesian activation estimation. For R users, the *fmri* library (cran.r-project.org/web/packages/fmri/index.html) contains functions for single-subject modeling under the GLM framework and computation of SPMs, to which adaptive smoothing algorithms and GRF theory can subsequently be applied. The package can perform signal detrending, but does not have further preprocessing capabilities.

A variety of software libraries for multivariate analysis methods are also available. For Matlab, the *GIFT* toolbox (mialab.mrn.org/software/gift/) includes routines for different variants of single- and multiple-subject ICA (InfoMax, FastICA, equivariant robust ICA, etc.), as well as for PCA. Alternatively, PCA capabilities are offered by the Matlab *FMRISTAT* package (www.math.mcgill.ca/

keith/fmristat/), while ICA algorithms can be found in *MELODIC* (www.fmrib.ox. ac.uk/fsl/melodic/index.html), a toolbox belonging to the *FSL* library, and in two R packages: the above mentioned *fmri* and *AnalyzeFMRI* (cran.r-project.org/web/ packages/AnalyzeFMRI/index.html). Both *MELODIC* and *fmri* can perform spatial and temporal data decomposition, whereas *AnalyzeFMRI* only focuses on the spatial variant.

PyMVPA (www.pymvpa.org) is by far the most comprehensive piece of software for machine-learning analysis of fMRI data. *PyMVPA* is written in Python and has many functionalities for feature selection (for example, recursive feature elimination and several ranking methods), classification (including SVM and logistic regression called via wrappers from external computational tools), and cross-validation (such as leave-one-out and bootstrapping procedures). Other popular machine-learning-oriented packages are *MVPA* (code.google.com/p/princeton-mvpa-toolbox/), *3dsvm* (lacontelab.org/3dsvm.htm), and *FACT* (sites.google.com/site/chuanglab/software/ fact). *MVPA*, implemented in Matlab, focuses on pattern classification analysis and includes routines for cross-validation and synthetic data generation; *3dsvm* is a command-line plug-in of *AFNI* for SVM classification; the stand-alone software *FACT* offers, among others, several routines for temporal clustering (for example, *fuzzy k*-means and Kohonen clustering neural networks).

References

1. K.K. Kwong, J.W. Belliveau, D.A. Chesler, I.E. Goldberg, R.M. Weissko, B.P. Poncelet, D.N. Kennedy, B.E. Hoppel, M.S. Cohen, R. Turner, H.M. Cheng, T.J. Brady, B.R. Rosen, Dynamic magnetic resonance imaging of human brain activity during primary sensory stimulation. Proc. Natl. Acad. Sci. (USA) **89**, 5675–5679 (1992)
2. N. Logothetis, A. Wandell, Interpreting the bold signal. Ann. Rev. Physiol. **66**, 735–769 (2004)
3. P. Belin, R. Zatorre, R. Hoge, A. Evans, B. Pike, Event-related fmri of the auditory cortex. NeuroImage **10**, 417–429 (1999)
4. R. Buckner, J. Goodman, M. Burock, M. Rotte, W. Koutstaal, D. Schacter, B. Rosen, A. Dale, Functional-anatomic correlates of object priming in humans revealed by rapid presentation event-related fmri. Neuron **20**, 285–296 (1998)
5. K. Ochsner, A. Bunge, J. Gross, J. Gabrieli, Rethinking feelings: An fmri study of the cognitive regulation of emotion. J. Neurosci. **14**, 1215–1229 (2002)
6. E. Falk, E. Berkman, T. Mann, B. Harrison, M. Lieberman, Predicting persuasion-induced behavior change from the brain. J. Neurosci. **30**, 8421–8424 (2010)
7. A. Sanfey, J. Rilling, J. Aronson, L. Nystrom, J. Cohen, The neural basis of economic decision-making in the ultimatum game. Science **300**, 1755–1758 (2003)
8. J. Sepulcre, H. Liu, T. Talukdar, I. Martincorena, T. Yeo, R. Buckner, The organization of local and distant functional connectivity in the human brain. PLoS Comput. Biol. **6**, e1000808 (2010)
9. H. Whalley, E. Simonotto, S. Flett, I. Marshall, K. Ebmeier, D. Owens, N. Goddard, E. Johnstone, S. Lawrie, fmri correlates of state and trait effects in subjects at genetically enhanced risk of schizophrenia. Brain **127**, 478–490 (2004)
10. G. Honey, E. Pomarol-Clotet, P. Corlett, R. Honey, P. McKenna, E. Bullmore, P. Fletcher, Functional dysconnectivity in schizophrenia associated with attentional modulation of motor function. Brain **128**, 2597–2611 (2005)

11. M. Wengenroth, M. Blatow, J. Guenther, M. Akbar, V. Tronnier, C. Stippich, Diagnostic benefits of presurgical fmri in patients with brain tumours in the primary sensorimotor cortex. Eur. Radiol. **21**, 1517–1525 (2011)
12. R. Marshall, E. Zarahn, L. Alon, B. Minzer, R. Lazar, J. Krakauer, Early imaging correlates of subsequent motor recovery after stroke. Ann. Neurol. **65**, 596–602 (2009)
13. R. Wise, I. Tracey, The role of fmri in drug discovery. J. Magn. Reson. Imag. **23**, 862–876 (2006)
14. D. Borsook, L. Becerra, R. Hargreaves, A role for fmri in optimizing CNS drug development. Nat. Rev. Drug Discov. **5**, 411–424 (2006)
15. M. Lindquist, The statistical analysis of fmri data. Stat. Sci. **23**, 439–464 (2008)
16. E. Amaro Jr., G. Barker, Study design in fmri: Basic principles. Brain Cognit. **60**, 220–232 (2006)
17. W. Machielsen, S. Rombouts, F. Barkhof, P. Scheltens, M. Witter, fmri of visual encoding: Reproducibility of activation. Hum. Brain Mapp. **9**, 156–164 (2000)
18. K. Friston, E. Zarahn, O. Josephs, R. Henson, A. Dale, Stochastic designs in event-related fmri. NeuroImage **10**, 607–619 (1999)
19. O. Josephs, R. Turner, K. Friston, Event-related fmri. human brain mapping. Hum. Brain Mapp. **9**, 243–257 (1997)
20. R. Buxton, K. Uludag, D. Dubowitz, T. Liu, Modeling the hemodynamic response to brain activation. NeuroImage **23**, S220–S233 (2004)
21. D. Donaldson, S. Petersen, J. Ollinger, R. Buckner, Dissociating state and item components of recognition memory using fmri. NeuroImage **13**, 129–142 (2001)
22. M. Greicius, K. Supekar, V. Menon, R. Dougherty, Resting-state functional connectivity reflects structural connectivity in the default mode network. Cerebr. Cortex **19**, 72–78 (2009)
23. J.V. Hajnal, R. Myers, A. Oatridge, J.E. Schwieso, I.R. Young, G.M. Bydder, Artifacts due to stimulus correlated motion in functional imaging of the brain. Magn. Reson. Med. **31**, 283–291 (1994)
24. S. Hamdy, D. Mikulis, A. Crawley, S. Xue, H. Lau, S. Henry, N. Diamant, Identifying global anatomical differences: Deformation-based morphometry. Am. J. Phisiol. **277**, G219–G225 (1999)
25. T. Stephan, E. Marx, H. Bruckmann, T. Brandt, M. Dieterich, Lid closure mimics head movement in fmri. Neuroimage **16**, 1156–1158 (2002)
26. D. Abbott, H. Opdam, R. Briellman, G. Jackson, Brief breath holding may confound functional magnetic resonance imaging studies. Hum. Brain Mapp. **24**, 284–290 (2005)
27. X. Hu, T.H. Le, T. Parrish, P. Erhard, Retrospective estimation and correction of physiological fluctuation in functional mri. Magn. Reson. Med. **34**, 201–212 (1995)
28. A. Moelker, P.M.T. Pattynama, Acoustic noise concerns in functional magnetic resonance imaging. Hum. Brain Mapp. **20**, 123–141 (2003)
29. A. Gordon, R. Smith, K. Keramatian, B. Luus, A. Weinberg, J. Smallwood, J. Schooler, K. Christoff, Mind-wandering, awareness, and task performance: An fmri study. Can. J. Exp. Psychol. **61**, 210–216 (2007)
30. J. Ashburner, C. Hutton, R. Frackowiak, I. Johnsrude, C. Price, K. Friston, Identifying global anatomical differences: Deformation-based morphometry. Hum. Brain Mapp. **6**, 348–357 (1998)
31. S.M. Smith, in *Preparing fmri Data for Statistical Analysis*, ed. by P. Jezzard, P.M. Matthews, S.M. Smith. Functional MRI: An Introduction to Methods (Oxford University Press, Oxford, 2001)
32. J. Tanabe, D. Miller, J. Tregellas, R. Freedman, F.G. Meyer, Comparison of detrending methods for optimal fmri preprocessing. NeuroImage **15**, 902–907 (2002)
33. M.J. Brammer, in *Head Motion and Its Correction*, ed. by P. Jezzard, P.M. Matthews, S.M. Smith. Functional MRI: An Introduction to Methods (Oxford University Press, Oxford, 2001)
34. L. Freire, J.F. Mangin, Motion correction algorithms may create spurious brain activations in the absence of subject motion. NeuroImage **14**, 709–722 (2001)

35. G. Glover, T.-Q. Li, D. Ress, Image-based method for retrospective correction of physiological motion effects in fmri: Retroicor. Magn. Reson. Med. **44**, 162–167 (2000)
36. K.-H. Chuang, J.-H. Chen, Impact: Image-based physiological artifacts estimation and correction technique for functional mri. Magn. Reson. Med. **46**, 344–353 (2000)
37. F. Crivello, T. Schormann, N. Tzourio-Mazoyer, P. Roland, K. Zilles, B. Mazoyer, Comparison of spatial normalization procedures and their impact on functional maps. Hum. Brain Mapp. **16**, 228–250 (2002)
38. J. Talairach, P. Tournoux, *Co-Planar Stereotaxic Atlas of the Human Brain* (Thieme, New York, 1988)
39. A. Evans, D. Collins, S. Mills, E. Brown, L. Kelly, T. Peters, *3D* statistical neuroanatomical models from 305 *MRI* volumes. Proceedings of the IEEE Nuclear Science Symposium and Medical Imaging Conference, vol. 3, pp. 1813–1817 (1993)
40. P. Fransson, K.-D. an Merboldt, K.M. Petersson, M. Ingvar, J. Frahm, On the effects of spatial filtering: A comparative fmri study of episodic memory encoding at high and low resolution. NeuroImage **16**, 977–984 (2002)
41. G. Aguirre, E. Zarahn, M. D'Esposito, The variability of human, bold hemodynamic responses. NeuroImage **8**, 360–369 (1998)
42. R. Menon, S. Ogawa, J. Strupp, P. Andersen, K. Ugurbil, Bold based functional mri at 4 tesla includes a capillary bed contribution: Echo-planar imaging mirrors previous optical imaging using intrinsic signals. Magn. Reson. Med. **33**, 453–459 (1995)
43. G. Glover, Deconvolution of impulse response in event-related bold fmri. NeuroImage **9**, 416–129 (1999)
44. J. Rajapske, F. Kruggel, J. Maisog, D. Von Cramon, Modeling hemodynamic response for analysis of functional mri time-series. Hum. Brain Mapp. **6**, 283–300 (1998)
45. N. Lazar, *The Statistical Analysis of Functional MRI Data* (Springer, New York, 2008)
46. T. Hastie, R. Tibshirani, J. Friedman, *The Elements of Statistical Learning: Data Mining, Inference and Prediction* (Springer, New York, 2009)
47. T. Mitchell, R. Hutchinson, R.S. Niculescu, F. Pereira, X. Wang, M. Just, S. Newman, Learning to decode cognitive states from brain images. Mach. Learn. **57**, 145–175 (2004)
48. F. Pereira, T. Mitchell, M. Botvinick, Machine learning classifiers and fmri: A tutorial overview. NeuroImage **45**, S199–S209 (2009)
49. F. Pereira, G. Gordon, The Support Vector Decomposition Machine. Proceedings of the International Conference on Machine Learning (ICML) (2006)
50. V. Calhoun, T. Adali, G. Pearlson, J. Pekar, Spatial and temporal independent component analysis of functional mri data containing a pair of task-related waveforms. Hum. Brain Mapp. **13**, 43–53 (2001)
51. Y. Shimizu, M. Barth, C. Windischberger, E. Moser, S. Thurner, Wavelet-based multifractal analysis of fmri time series. Neuroimage **22**, 1195–1202 (2004)
52. C. Neil, H. Trevor, J. Iain, Statistical models for image sequences. Technical report, Stanford University (1998)
53. N. Lange, S. Zeger, Non-linear fourier time series analysis for human brain mapping by functional magnetic resonance imaging. J. Roy. Stat. Soc. C (Appl. Stat.) **46**, 1–29 (1997)
54. M. Misaki, Y. Kim, P. Bandettini, N. Kriegeskorte, Comparison of multivariate classifiers and response normalizations for pattern-information fmri. Neuroimage **53**, 103–118 (2010)
55. J. Haxby, I. Gobbini, M. Furey, A. Ishai, J. Schouten, P. Pietrini, Distributed and overlapping representations of faces and objects in ventral temporal cortex. Science **293**, 2425–2430 (2001)
56. N. Kriegeskorte, R. Goebel, P. Bandettini, Information-based functional brain mapping. Proc. Natl. Acad. Sci. U.S.A. **103**, 3863–3868 (2006)
57. L.K. Hansen, J. Larsen, F.A. Nielsen, S.C. Strother, E. Rostrup, R. Savoy, N. Lange, J. Sidtis, C. Svarer, O.B. Paulson, Generalizable patterns in neuroimaging: How many principal components? NeuroImage **9**, 534–544 (1999)
58. F. De Martino, F. Gentile, F. Esposito, M. Balsi, F. Di Salle, R. Goebel, E. Formisano, Classification of fmri independent components using ic-fingerprints and support vector machine classifiers. NeuroImage **34**, 177–194 (2007)

59. K. Norman, S. Polyn, G. Detre, J. Haxby, Beyond mind-reading: multi-voxel pattern analysis of fmri data. Trends Cognit. Sci. **10**, 424–430 (2006)
60. S. LaConte, S. Peltier, X. Hu, Real-time fmri using brain-state classification. Hum. Brain Mapp. **28**, 1033–1044 (2007)
61. D. Cox, L. Savoy, Functional magnetic resonance imaging (fmri) 'brain reading': detecting and classifying distributed patterns of fmri activity in human visual cortex. Neuroimage **19**, 261–270 (2003)
62. X. Wang, R. Hutchinson, T. Mitchell, Training *fMRI* classifiers to detect cognitive states across multiple human subjects. Proceedings of the Conference on Neural Information Processing Systems (2003)
63. S. LaConte, S. Strother, V. Cherkassky, J. Anderson, X. Hu, Support vector machines for temporal classification of block design fmri data. NeuroImage **26**, 317–329 (2005)
64. T. Mitchell, R. Hutchinson, M. Just, R. Niculescu, F. Pereira, X. Wang, Classifying instantaneous cognitive states from *fMRI* data. AMIA Annual Symposium Proceedings, pp. 465–469 (2003)
65. C. Cortes, V. Vapnik, Support vector networks. Mach. Learn. **20**, 273–297 (1995)
66. S. Ryali, K. Supekar, D.A. Abrams, V. Menon, Sparse logistic regression for whole-brain classification of fmri data. Neuroimage **51**, 752–764 (2010)
67. C. Baudelet, B. Gallez, Cluster analysis of bold fmri time series in tumors to study the heterogeneity of hemodynamic response to treatment. Magn. Reson. Med. **49**, 135–145 (2003)
68. J. Lancaster, M. Woldorff, L. Parsons, M. Liotti, C. Freitas, L. Rainey, P. Kochunov, D. Nickerson, S. Mikiten, P. Fox, Automated talairach atlas labels for functional brain mapping. Hum. Brain Mapp. **10**, 120–131 (2000)
69. R. Heller, D. Stanley, D. Yekutieli, N. Rubin, Y. Benjaminia, Cluster-based analysis of fmri data. Neuroimage **33**, 599–608 (2006)
70. C. Goutte, P. Toft, E. Rostrup, F.A. Nielsen, K.L. Hansen, On clustering fmri time series. Neuroimage **9**, 298–310 (1999)
71. J. Ye, N. Lazar, Y. Li, Geostatistical analysis in clustering fmri time series. Stat. Med. **28**, 2490–2508 (2009)
72. D. Balslev, F.A. Nielsen, S.A. Frutiger, J.J. Sidtis, T.B. Christiansen, C. Svarer, S.C. Strother, D.A. Rottenberg, L.K. Hansen, O.B. Paulson, I. Law, Cluster analysis of activity-time series in motor learning. Hum. Brain Mapp. **15**, 135–145 (2002)
73. M.J. Fadili, S. Ruan, D. Bloyet, B. Mazoyer, A multistep unsupervised fuzzy clustering analysis of fmri time series. Hum. Brain Mapp. **10**, 160–178 (2000)
74. L. Stanberry, R. Nandy, D. Cordes, Cluster analysis of fmri data using dendrogram sharpening. Hum. Brain Mapp. **20**, 201–219 (2003)
75. J. MacQueen, Some methods for classification and analysis of multivariate Observations. Proceedings of 5th Berkeley Symposium on Mathematical Statistics and Probability (University of California Press, CA, 1967), pp. 281–297
76. J. Bezdek, R. Ehrlich, W. Full, Fcm: The fuzzy c-means clustering algorithm. Comp. Geosci. **10**, 191–203 (1984)
77. B. Yeo, W. Ou, Clustering fmri time series(2004), http://people.csail.mit.edu/ythomas/ unpublished/6867fMRI.pdf. Accessed 20 June 2012
78. P. Filzmoser, R. Baumgartner, E. Moser, A hierarchical clustering method for analyzing functional mr images. Magn. Reson. Imag. **17**, 817–826 (1999)
79. R. Baumgartner, C. Windischberger, E. Moser, Quantification in functional magnetic resonance imaging: Fuzzy clustering vs. correlation analysis. Magn. Reson. Imag. **16**, 115–125 (1998)

Sensory Neuroprostheses: From Signal Processing and Coding to Neural Plasticity in the Central Nervous System

Fivos Panetsos, Abel Sanchez-Jimenez, and Celia Herrera-Rincon

Abstract To develop neuroprostheses that will provide the nervous system with artificial sensory input through the sensory nerves to which they will be connected, on one hand we have to determine how external stimuli are represented, coded and transmitted by the Nervous System, how neurons and neuronal ensembles process, encode and transmit perceptual information. On the other we need to know how the central nervous system reacts to the implanted neuroprostheses and quantify its anatomic and functional alterations due to the artificial input it receives from our devices. Here we present mathematical and electrophysiological methods for signal acquisition, analysis, and information coding in the tactile sensory system that include a wavelet and principal component analysis-based method for neural signal analysis and different types of frequency-based signal processing and coding performed simultaneously by the sensory neurons. Finally we present a quantitative morphological study of the effects of the neuroprosthetic stimulation using a stereological approach.

F. Panetsos • C. Herrera-Rincon (✉)
Neurocomputing and Neurorobotics Research Group and Department of Applied Mathematics (Biomathematics), Complutense University of Madrid, Avda Arcos de Jalon 118, 28037 Madrid, Spain

School of Optics, Complutense University of Madrid, Avda Arcos de Jalon 118, 28037 Madrid, Spain
e-mail: fivos.panetsos@opt.ucm.es, celia.herrer@opt.ucm.es

A.S.-Jimenez
Neurocomputing and Neurorobotics Research Group and Department of Applied Mathematics (Biomathematics), Complutense University of Madrid, Avda Arcos de Jalon 118, 28037 Madrid, Spain

Faculty of Biology, Complutense University of Madrid, Antonio Novais 12, 28040 Madrid, Spain
e-mail: abelsanchez@bio.ucm.es

P.M. Pardalos et al. (eds.), *Optimization and Data Analysis in Biomedical Informatics*, Fields Institute Communications 63, DOI 10.1007/978-1-4614-4133-5_8, © Springer Science+Business Media New York 2012

Mathematics Subject Classification (2010): Primary 92; Secondary 92C20

1 Introduction

Sensory neuroprostheses are designed to restore lost sensory functions by substituting the natural sensory input with another, artificial, but resembling the natural one. In chronically implanted amputee subjects, they record the electrical activity of artificial sensors placed in the periphery of the human body and transmit it (suitably coded as nerve activity) to the central nervous system through the sensory nerves to which they are connected. To achieve this objective we need theoretical knowledge and experimental tools to artificially stimulate the peripheral nerves in such a way that the elicited responses of the neural populations of the sensory systems will be similar to those generated by natural stimuli. Our specific objectives in this endeavor are:

1. To determine how external stimuli are represented, coded and transmitted by the Nervous System (NS) through the different processing stations of the sensory pathways, It is necessary to determine how interactions between sensory inputs and the activity of Central Nervous System (CNS) neurons create the internal representation of real-world stimuli in the different stations of the sensory systems.
2. To obtain useful knowledge on the relation between external stimuli, activity patterns of the afferent fibers and neural activity in different stations of the somatosensory pathway. It is important to determine the use that neurons and neuronal ensembles make of synchronous activity and oscillatory behavior to process, encode and transmit perceptual information on sensory stimuli.
3. To deliver artificial input to the brain by means of implanted neuroprosthesis and evaluate how CNS reacts and quantify its anatomic and functional alterations due to our manipulations.

   In the case of the somatosensory system our neuroprosthesis should be a bionic arm implanted to the stump of an amputee person and connected directly to the fibers of the peripheral nerve.

1.1 Principles for Signal Coding to Produce Perception

The somatosensory system receives and processes information from the body surface and from deep tissues and viscera. Different sensations result from external stimuli exciting a variety of somatic receptors that are distributed throughout the body. Sensory stimuli are encoded as patterns of electrical activity (action potentials) and transmitted to the CNS through the primary afferent fibers, which constitute the peripheral sensory nerves. Information on the external world is transmitted through the nerves as complex spatiotemporal and intensity patterns of electrical pulses and then analyzed and combined in subsequent processing stages to produce perception. The perception of the external world by the CNS is the result of a constant interplay between incoming signals and dynamic internal representations of the external world. It is long established that neurons encode incoming signals through a linear function between stimulus intensity and spiking frequency. However, encoding is not performed only in such a simple and inefficient way. A great amount of recent data suggests that more complex and efficient procedures like synchronized activity, interspike intervals (high-order statistics) and oscillating behavior of neural cells play an important role in information processing, whether sensory, motor or mental. Unfortunately, our knowledge on these procedures is too limited to allow us to define protocols of artificial stimulation of the sensory nerves that could be appropriately interpreted by the CNS.

In order to create an artificial device that faithfully reproduces peripheral sensory signals it is therefore necessary to determine in detail, in first place, the fundamental principles that support the coding of sensory stimuli. This condition, however, is not sufficient. We must be able, as well, to reproduce, through electrical stimulation of the nerve fibers, the spatiotemporal pattern of activity generated by the physiological sensory stimuli applied to the skin. This enterprise, however, is limited by a number of biological facts, which are the source of serious obstacles:

- In humans, the sensory nerves convey thousands of nerve fibers (axons), which are peripheral branches of the primary sensory neurons located in spinal, or dorsal root, ganglia. If artificial sensors are to be directly connected to the nerves there is a limit to the number of fibers that can be interfaced -from some tens to a few hundreds. The reasons for it are both physiological (neuronal and axonal degeneration, or limited biocompatibility) and technical (limitations in the size and number of electrode contacts).
- Each intact axon conveys a specific kind of sensory information from specific receptors (sensory submodalities as light touch, vibration, warmth...) towards neurons which process that information and eventually integrate it with similar -or different- information arriving from other afferent axons. The interfacing of artificial sensors to a sensory nerve, however, will submit to axons information of different sensory submodalities in an entirely unpredictable fashion.
- Nerve fibers are spatially arranged so that the peripheral receptor sheet (e.g., the skin) is mapped through the nerve onto the neurons of CNS (somatotopy). The implant of an electronic interface in a cut nerve cannot maintain this spatial organization substantially changing the somatotopic map. These maps are capable to partially reorganize, however, though the extent and shape of this reorganization is contingent upon different, not fully understood, variables.

Neurons receive information through their synapses that generate small local de- or hyper-polarizations of the neural membrane (excitatory post synaptic potentials -EPSPs and inhibitory post synaptic potentials -IPSPs respectively). These fluctuations of the membrane potential are integrated and transmitted along the membrane of the dendrites and the cell body. To be transmitted to other neurons information

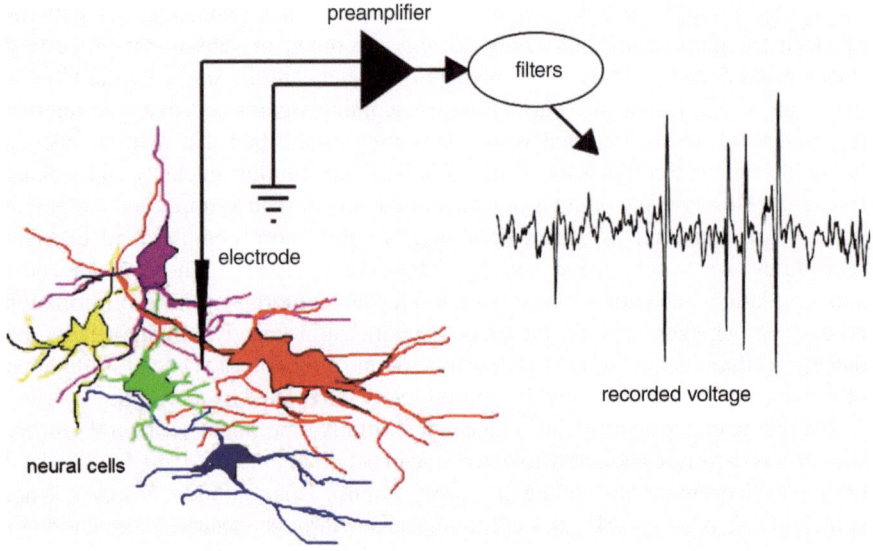

Fig. 1 Data acquisition from electrophysiological recordings: a tungsten electrode is placed into the neural tissue at some micrometer from neural cells (somata, axons, dendrites). Any modification of the membrane potential is recorded extracellularly from the ionic medium, amplified, filtered and then stored and visualized. Rapid voltage changes (*spikes*) and noisy background are clearly recognized in the record

is then transcripted to a digital code as pulses of the same duration and amplitude named spikes (0.2–2.0 ms, 80 mV, Fig. 1, right). Information coding is mainly based on the number of spikes, their frequency and their temporal distribution. In addition, groups of neurons process and transmit information in a population manner using a, generally unknown, distributed code.

1.2 Signal Acquisition and Processing

In-vivo simultaneous monitoring of firing patterns of many neurons and nerve fibers is achieved by placing extracellular electrodes to the brain without significant tissue damage. The basic electrical circuit used in such an experiment is shown in Fig. 1. The circuit amplifies the potential between the ground (usually measured by placing a wire under the scalp) and the tip of the microelectrode. The potential measured at the tip of the electrode reflects the current flow in the extracellular medium due to the action potentials generated by neurons and axons near it. Extracellular electrodes normally record the activity of many nearby neurons with a consequent difficulty to differentiate the spikes generated by these neurons. In addition, local field potentials, thought to be generated from the current flow into the dendrites of the neurons, have a sufficiently low bandwidth and can be filtered out from the action potentials.

Spikes recorded from different neurons differ in shape and amplitude that depends on the distance between neuron and electrode tip. This gives us a chance to assign spikes to different cells. The spikes from a single cell will form clusters in high-dimensional feature space [8,9,21] after a process of classification called spike sorting. An observation of the trace in Fig. 1 suggests the existence of different action potentials and a significant amount of background noise. For the reconstruction of the underlying neural activity the general assumption is that each neuron produces a different, reproducible waveform, which is then contaminated by an additive noise, so the Signal to Noise Ratio (SNR) can be rather low [20, 31]. Sources for noise spam from Johnson noise in the electrode and electronics, background activity of distant neurons [8,9] to electrode micro-movement [32] passing through waveform misalignment [17], and the variation of the action potential shape as a function of recent firing history [8, 26]. Additionally, spikes proceeding from different cells may overlap. The problem of automatically classifying the different shapes can be addressed either in the context of the full time-sampled spike-shape or of a reduced feature set, such as the principal components [10, 11], or a wavelet basis [16]. Simultaneous monitoring of the same cell intra and extracellularly with controlled spike sorting, shows that manual sorting by professional operators with using tetrode recording fails between 10 and 30% of the cases while single electrode recordings perform even worst (up to 50%, [14]). In a 1-day typical experiment we can easily have 10^4–10^5 spikes per recording electrode, hence optimal automatic separation techniques are necessary.

To obtain an optimal performance in spike identification we developed a novel method for extraction of spike features, the Wavelet Shape-Accounting Classifier (WSAC, [24]), based on a combination of principal component analysis (PCA) and continuous wavelet transform (WT). The method automatically tunes its WT part to the data structure making use of knowledge obtained by PCA. In the PCA approach a set of orthogonal eigenvectors of the covariance matrix of the spike waveforms is estimated, then each spike is completely represented by a sum of the principal component vectors with the corresponding scale factors (scores) considered as spike features for sorting (Fig. 2, top-left). In the WT approach the coefficients of the WT

$$C(a,b) = \frac{1}{\sqrt{a}} \int_{-\infty}^{+\infty} s(t)\psi_{a,b}(t)dt, \quad \psi_{a,b}(t) = \psi\left(\frac{t-b}{a}\right) \quad (1)$$

are used to represent spikes in manner similar to a Fourier transform [16, 19, 27].

In WSAC we first look for representative spike WaveForms (rWFs). With real data with a PCA approach we normally obtain partially overlapping clouds of spikes of different neurons represented in the PCA space. We localize the positions of spike density maxima and we average spike waveforms in a small neighborhood of each cloud center to obtain the mean or rWFs. Then we search for a set of wavelet parameters (a^*, b^*) maximizing the distance $|CrWF_i(a,b) - CrWF_j(a,b)|$ between the rWFs in the wavelet space. Finally we evaluate the coefficients of the wavelet for the

Fig. 2 *Top*: Typical spike
classification by PCA
resulting in overlapping
clouds corresponding to the
spikes of two probably
different neurons (*left*) and
histogram of spike
distribution along the PCA
space (*right*). *Bottom*:
Classification improves when
the WSAC approach is
employed as can be shown in
the less overlapping clouds
(*left*) and the more separated
distributions in the right
histogram (after Pavlov et al.
2007)

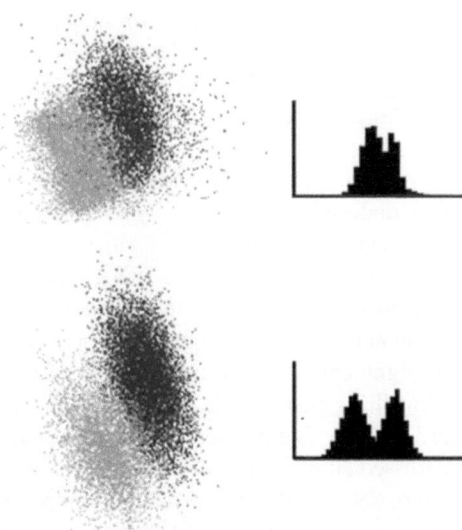

parameter sets we found for all spikes, $C_i(a^*, b^*)$. Now the clouds corresponding to
the identified neurons are better delimited and less overlapping in the wavelet than
in the PCA space (Fig. 2, bottom).

2 Neural Processing and Coding in the Tactile System

2.1 Experimental Model

Our first approach is to deliver batteries of tactile stimuli to the receptors of the
animal, record the responses of the sensory neurons to these stimuli and analyze
them trying to decipher how they are processing and coding the signals they receive.

Most of our experiments were performed in the trigeminal system of the rat
due to its highly topographic organization and the wide knowledge on its anatomy,
physiology and plasticity [7, 23, 33]. The trigeminal tactile system is principally
composed by a matrix of 33 long hairs (macrovibrissae, or main whiskers) located
on the snout of the animal with which the rat touches and discriminates external
objects similarly to the humans' fingers. Tactile information from these sensors is
conveyed to the CNS through the trigeminal nerve in a spatially organized, rapid
and direct manner (Fig. 3, left). This nerve enters the brain and synapse with the
neurons of the first relay station, the ipsilateral sensory trigeminal nuclei of the
brainstem. Neurons in these nuclei send axons that cross the midline and synapse
with the neurons of the contralateral somatosensory thalamus, that in turn project
to the primary somatosensory cortex. In all relay stations, information processing is
done by groups of neurons that are spatially distributed in the same manner as their

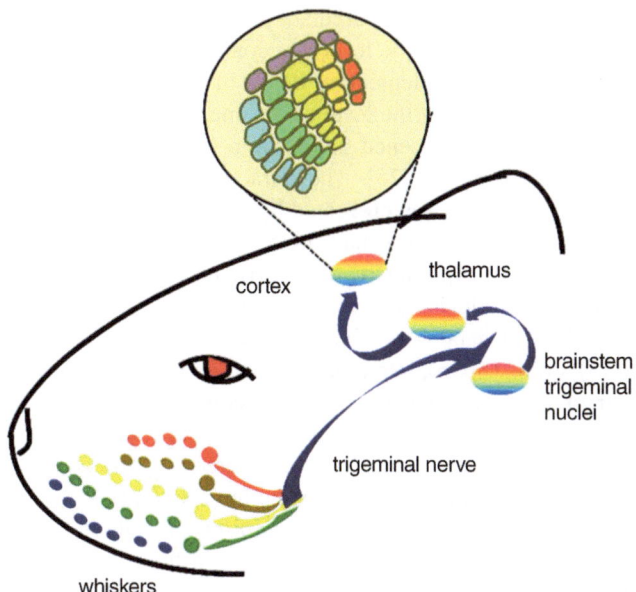

Fig. 3 The rat trigeminal system. A matrix of 33 long hairs (whiskers or vibrissae) on the snout of the animal forms the tactile apparatus equivalent to the humans' fingers. A high number of sensors in the base of every whisker transform tactile stimuli to electrical signals that are conveyed through the trigeminal nerve to the first relay station of the CNS, the sensory trigeminal nuclei of the brainstem (nuclei principalis, oralis, interpolaris and caudalis). Neurons in these nuclei send their axons to the contralateral thalamus, that in turn project to the primary somatosensory cortex. In all relay stations there are maps of the whiskers, formed by groups of neurons (somatotopic maps) each of which mainly process information from its corresponding vibrissa on the snout

corresponding whiskers on the snout of the rat (somatotopic maps). The main input of each of these groups comes from a single vibrissa on the snout. This arrangement can be detected by a number of histological and neurochemical markers (see also Fig. 6). In cortical layer IV these groups of neurons take the shape of barrels and form the cortical "barrel field" [37, 38].

Tactile stimulation and recordings of the neural responses The distal portions (i.e., free ends) of the vibrissae were stimulated using air-jets generated by a pneumatic pressure pump (10 psi, Picospritzer III, Parker Institute, Texas, USA) and delivered in a rostrocaudal direction. These stimuli allow the whiskers to move and vibrate in a natural fashion resembling contact with an object in a rostrocaudal direction [30]. Spontaneous activity of the neural cells was recorded for 180 s, and a sequence of 50 pulses of 100 ms duration was applied at 1 Hz. Spontaneous activity was recorded for additional 120 s, and then the following protocol was used: 14-ms duration air puffs in five-second long pulse trains were presented at 1, 2, 3, 5, 8, 10, 12, 15, 20, 25, 30, 35, and 40 Hz. Each frequency was presented ten times in a random order with a 3 s interval between trains, and the experiment was terminated with a final sequence of 50 pulses of 100 ms at 1 Hz.

Animals were anesthetized with urethane or ketamine+hydrochloride/xylazine adult albino Wistar rats. Anesthetic level was controlled by means of an electroencephalogram (EEG), and supplementary doses of anesthetic (10% of the initial dose) were administrated when the level of the anesthesia was lowered. The scalp was removed, the bone was opened 2.0–3.5 mm lateral to the midline and 8.0–12.5 mm posterior to Bregma [25]. The dura was removed, and drying of the exposed surface of the brain was prevented by covering it with vaseline oil. After the removal of the scalp, a hole was made in the frontal part of the skull to insert the EEG macroelectrode. All experiments were carried out according to EU Directives (86/609/EC) and national legislation (R.D. 1201/2005) on this matter.

Electrophysiological data were obtained from the sensory trigeminal nuclei, principalis (Pr5), oralis (Sp5o), interpolaris (Sp5i) and caudalis (Sp5c) that receive in parallel direct information from the ipsilateral whiskers (Fig. 3) using 0.8–2.0 $M\Omega$ tungsten microelectrodes. As the electrode was driven vertically into the brainstem, the vibrissae were manually stimulated under microscope observation using a thin brush until neural responses were obtained. Once a good response was obtained, the receptive field was manually determined, and the vibrissa that elicited the maximum activation was labeled as the centre of the receptive field (principal vibrissa). The principal vibrissa was then stimulated using the protocol described below. Recorded signals were amplified, filtered (0.3–3 KHz) online and digitalized (300 Hz EEG recordings, 20 KHz extracellular recordings) for storage.

2.2 Data Analysis

A first analysis is performed by means of peristimulus histograms (PSTHs) with a time resolution of 1 ms. PSTHs are widely used for the analysis of neural responses, but their power is quite limited when the neural dynamics under observation are complex and/or go beyond variations in the mean spiking frequencies. For this reason we used functions like early and global behavior, latency changes, and temporal consistency of the responses for stimulation frequencies in the range of 1–40 Hz [30].

The early behavior of the neural response, the part of the response contained in a short time interval after the onset of the stimulus, is measured by the Repetition Rate Transfer Function (RRTF). This function compares the first stimulus of the 5s train with subsequent stimuli at each stimulation frequency. The spikes generated by each deflection of the vibrissa in this time interval from the second stimulus to the end of the train are averaged and then divided by the spikes evoked by the first stimulus of the train in the same time interval. An RRTF value greater than 1 indicates an increased spike rate (potentiation); a value less than 1 indicates a decrease in the response rate (adaptation).

The global behavior of the neural response measured by the Total Spike Rate (TSR), represents the total number of spikes evoked over an extended stimulation period at each stimulation frequency, 5s in our case, since the integration over multiple stimuli is believed to be crucial for frequency discrimination in primates [28, 29]. To compare different recordings and different nuclei responses, we normalized TSR values along the 1–40 Hz range from each recording by the value of the frequency with the highest TSR.

Response latency at each stimulus frequency was calculated as the average time between stimulus onset and the appearance of the first spike in each train of air-jet stimuli. This parameter was evaluated via the average cycle histogram and defined as the post-stimulus time at which the response amplitude reached 50% of its peak value. To properly compare the different recordings, response latencies were normalized with respect to the maximum value among stimulation frequencies. We studied the temporal consistency of spike timing across stimuli cycles by considering the i-th spike as a vector of unitary length and argument $\theta_i = 2\pi \left(\frac{t_i}{T}\right)$, $0 \leq \theta_i \leq 2\pi$, and measuring the phase-locking of the responses to the external stimuli by means of the Vector Strength (VS) function [12]:

$$VS = \frac{\sqrt{\left(\sum_{i=1}^{n} \cos(\theta_i)\right)^2 + \left(\sum_{i=1}^{n} \sin(\theta_i)\right)^2}}{n} \tag{2}$$

where n is the total number of spikes evoked during the stimulus train, T is the period of the stimulus frequency, and ti is the time interval between the most recent vibrissa deflection and the i-th spike. VS takes values between 0 and 1, from random spiking to perfect phase locking.

2.3 Information Processing by the Sensory Neurons

In early behavior neural responses combine frequency-dependent adaptation expressed as low-pass filtering with potentiation at specific frequencies (band-pass) or clear adaptive/low-pass behavior, as shown by the neural responses in a short time interval after the onset of the stimulus (Fig. 4). Potentiation in Pr5 occurs mainly between 2 and 15 Hz whereas Sp5i recordings show potentiation mostly near 12 Hz. Significant intra- and internuclear differences are observed in band-pass and non band-pass behavior for both, percentages and mean RTFF values for each stimulation frequency. Band-pass RTFF values were significantly higher in both Pr5 and Sp5i for almost all frequencies [30].

In global behavior TSR values show potentiation with the stimulation frequency, and these potentiations are either simple (logarithmic high-pass behavior) or combined (high-pass with additional band-pass potentiations in the range of 8–20 Hz) without statistically significant differences between Pr5 and Sp5i (Fig. 5). Intranuclear comparisons of TSR values at each stimulation frequency between band-pass and non band-pass recordings show very significant differences in absolute and normalized values in both locations [30].

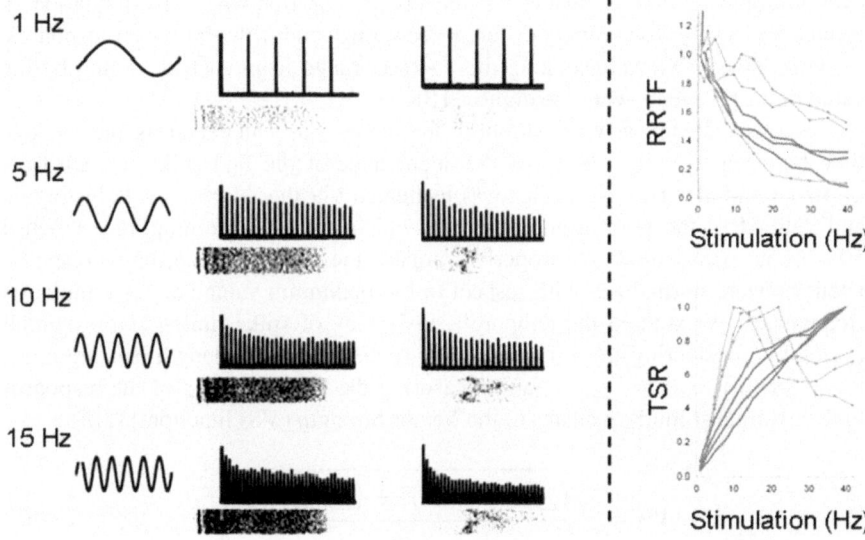

Fig. 4 *Left*: *First column*, frequency of whisker stimulation; *second* and *third columns*, responses of four neurons to the tactile stimuli of the whiskers at these frequencies (two are shown as peristimuli histograms and two as rasters) *Right*: graphic representation of the RRTF and TSR behavior of different neurons to 1–40 Hz stimuli. High pass, low pass and band-pass behavior are shown

Response latencies increase significantly as a function of stimulation frequency in 80.4% of Pr5 and 100% of Sp5i recordings. The increase of mean response latencies is quite slow, and statistically significant differences between the two nuclei arise after 15 Hz due to the more rapid increase of Sp5i values. Changes in the response latency have been proposed as a coding parameter in the vibrissae sensory pathway [1]. The temporal consistency of the neural responses is high in both nuclei and follows sigmoid curves. Between 1 and 5 Hz mean VS take values near 1.0 and decreases with the increasing of the stimulation frequency.

Early behavior is most likely related to specific properties of the stimulus (timing, number of involved vibrissae, relative velocity and direction of the object, etc.) while global behavior with long-lasting stimuli is probably used to encode information about the texture of the touched objects [28]. Mean firing rate, the TSR in our case, is considered the best candidate as neural code for tactile discrimination [18, 28, 29] and references therein). The total number of spikes fired in our 5 s stimulations increases with the increasing in stimulation frequency in both nuclei.

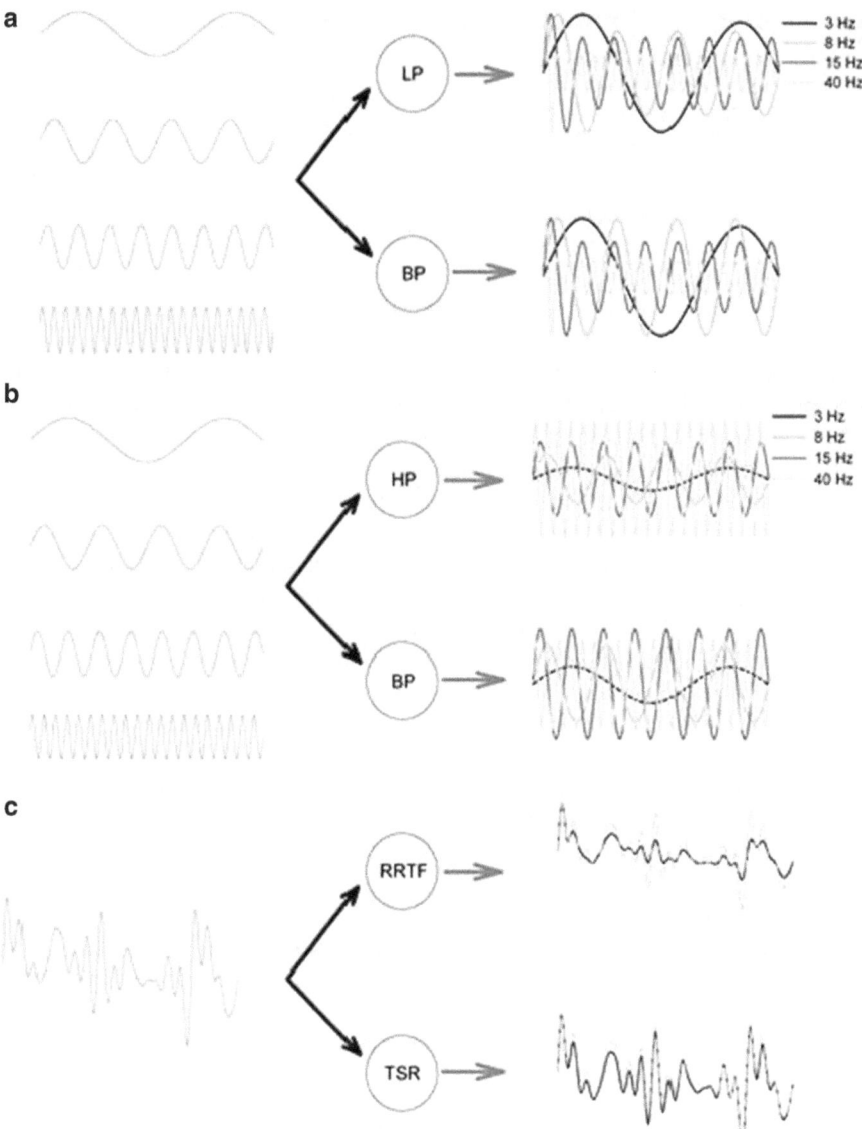

Fig. 5 Graphic representation of tactile signals (whisker displacement at 3–40 Hz) processing depending on the filtering properties of the sensory neurons. (**a**) Filtering of the early response (RFTF) by a low-pass and a band-pass (8 Hz) neuron. *Left*, whisker displacement; *right*, the incoming signal as "perceived" by the neuron. (**b**) The same as before but for the later response (TSR). (**c**) Signal processing and reconstruction of a complex táctiles signal (Leith) by a group of 1.000 low-pass, band-pass and high-pass neurons. The signal is "perceived" and reconstructed in a different manner in the RRTF and TSR spaces (*left*, reconstructed signal in dark)

3   Anatomic and Functional Alterations of the Brain Due to Neuroprosthetic Input

3.1   Experimental Approach

Current tools for morphological research allow us to see almost all the elements of an individual neuron and, by extension, of the whole nervous system. Thus, by morphological analysis we can identify and quantify the structural characteristics of individual neurons and / or complex regions of interest. Histological examination is able to detect microscopically from subcellular components (e.g., changes in size and shape of the nucleus, altered dendritic protein expression, etc.) to changes in the organization of the different functional regions (e.g., plastic cortex reorganization after learning or sensory deprivation, extent of damage after stroke, etc.) (see Fig. 6 for histological images at different levels of organization).

To evaluate the effects of the neuroprosthetic input to the CNS we took into account the symmetry of the body and that each side of the peripher is mainly connected to the contralateral cerebral cortex and that manipulations of a peripheral nerve should modify the contralateral cortex. Consequently, a comparison of the two cortices should give us an estimation of the effect of the unilateral experimental manipulation. For our study we used animals with complete unilateral transection of the trigeminal nerve, (Amputee or A-group), animals with the same manipulation but combined with chronic electrical stimulation of the transected nerve (Prosthetic or P-group) and intact animals (Control or C-group). C and A animals were left in their cages without further manipulation while P animals were held under 12 h/day continuous electrical stimulation (square pulses of 100 ms, 3.0 V, at 20 Hz) for 4 weeks. Stimulation parameters were selected from a wide range of previously tested stimulation protocols.

Cortical activity was evaluated histologically by estimating the metabolic activity of the cortical tissue and the volume of the active tissue in the barrel cortex. The metabolic activity of the neurons and in particular of their synapses is correlated to the expression of the cytochrome oxidase (CyO), a critical enzyme of the mitochondrial respiratory chain [34, 35], so by quantifying the quantity of CyO in the neural tissue and compare the two brain hemispheres we can evaluate the effect of our manipulation to the somatosensory cortex. The barrelettes of the brainstem, barreloids of the thalamus and barrels of the cortex are clearly distinguishable of underlying neuropil by their high enzymatic reactivity (see Fig. 3 and [36]). This allows us to accurately delineate areas according to their functional activity.

Traditionally, it has been possible to show certain morphological properties in the nervous system and make a qualitative or subjective interpretation of them. This qualitative analysis is still very useful today in the early stages of an investigation. However, the histological images obtained under a microscope represent a very small part of reality, so systematic random sampling is employed to provide

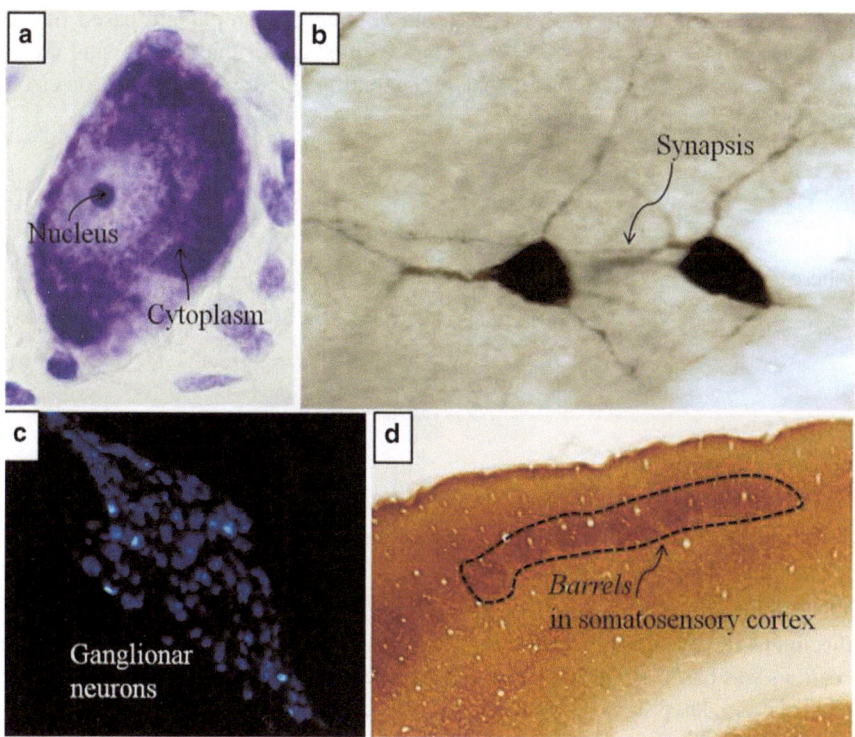

Fig. 6 Photomicrograhs of rat histological sections stained with different methods. (**a**) One Nissl-stained neuron showing basic components of cellular soma: nucleus and cytoplasm. In (**b**) two immunopositive neurons for a cytoplasmic protein (calbindin D-28k) are showed. These neurons are GABA-ergic interneurons located in the primary somatosensory cortex. (**c**) A neuronal agrupation or ganglion cells showing Fast-Blue (FB) positive sensory neurons. FB is a fluorescent retrograde neurotracer, what can be applied in the proximal stump of a transected peripheral nerve and traveling along the fibers reach and colour the neuronal body. (**d**) Barrel cortex stained for cytochrome-oxidase enzyme. *Dark zones* indicate high metabolic, and hence functional, activity. This histochemistry technique is very useful for studying cortical plasticity induced by manipulations of peripheral input (see text)

estimates with less variability and the need for more rigorous approaches led to the adaptation of mathematic tools for an objective quantitative analysis [4, 13].

The histological sampling is an essentially hierarchical procedure and must be done strictly from the first level. In the above example about the total number of neurons in the rat somatosensory cortex, including from the sacrifice of the animal to obtain a numerical data, there are many processes subject to bias and variability, whose error must also be measurable and quantified. Briefly, after sacrificing the animal, the brain is removed (three-dimensional object or block). The brain is cut into sections, on which study fields of view. Variability in brain samples is mainly due to the biological inter-individual variability (70%) and variability between blocks (20%), variability between sections, fields and measurements being limited

(5%, 3% and 2% respectively). Areas and volume estimations are based on the Cavalieri principle $\hat{V} = T \cdot \sum_{i=1}^{m} A_i$ applied in slices of thickness T and cross-sectioned area A_i estimated using the point counting [3,5,13,36]. Estimation errors are calculated as [6,13,22]:

$$CE(\hat{V})_{GJ} = \frac{1}{\sum A} \cdot \left(\frac{1}{12} \{3a + c - 4b\} \right)^{1/2} \tag{3}$$

where

$$a = \sum_{i=1}^{m} A_i \cdot A_i, \quad b = \sum_{i=1}^{m-1} A_i \cdot A_{i+1} \quad \text{and} \quad c = \sum_{i=1}^{m-2} A_i \cdot A_{i+2} \tag{4}$$

3.2   Quantitative Morphological Analysis in the Neuroprosthetical Approach

We performed an optical density analysis of the neural activity by determining the intensity of the CyO staining through gray-level measures. Selection of the sections was based on strict morphological criteria. Possible aberrations originating from the optical system of the camera were corrected by background subtraction. Multiple sections per animal were analyzed taking different samples from each slide, each sample consisting on the mean value of the pixels of a fixed-size window. Globally, Control- and Prosthetic- animals displayed statistically similar staining intensities in the two hemispheres (interhemispheric differences lower than 2.5%), while Amputated-animals presented significantly lower intensities in the manipulated cortex (approximately 9%, Fig. 7). Intergroup comparisons show statistical differences between C- and A- and P- and A- groups and no differences between C- and P-animals.

By applying the Cavalieri-point counting method, we estimated the volume of the highest enzymatic reactivity-areas, and therefore areas of high functional activity, in each trigeminal nucleus of the two hemispheres. The ratio of the estimated volume for the experimental hemisphere vs. control hemisphere of each animal was used as variable for intra-and inter-group comparisons. Estimations of the active volume in the somatosensory cortex were similar to the metabolic activity levels with the same statistical interhemispheric and intragroup differences [15].

Our main findings were that chronic deafferentation has important consequences on the structure of the CNS, reducing the functionally active structure in all nuclei of the somatosensory pathway. The artificial input causes the reorganization of the thalamic nucleus and somatosensory cortex, without significant consequences at lower levels. Thalamic and cortical activity, silenced after the amputation, was restored after the neuroprosthetical stimulation.

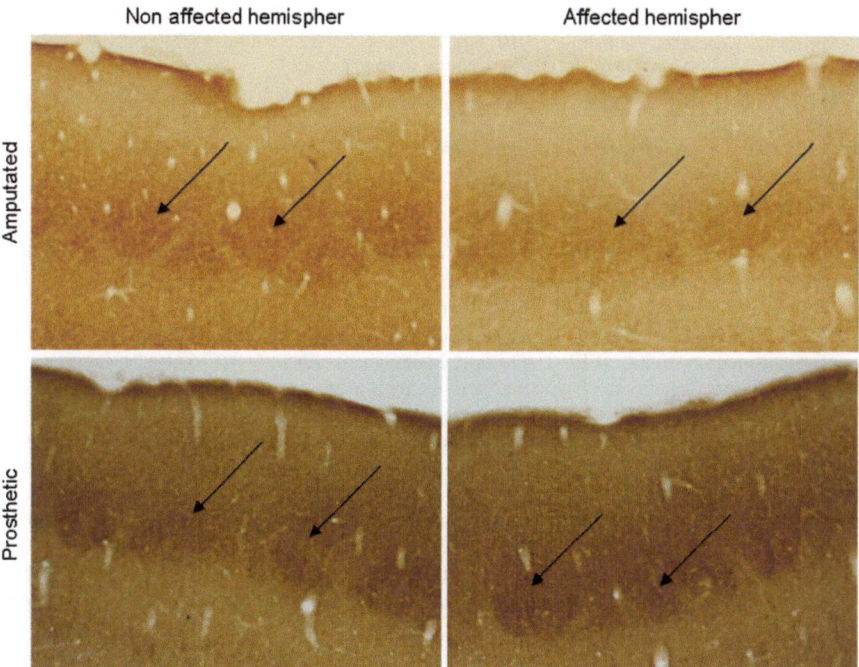

Fig. 7 Sections of the somatosensory cortex of amputated and prosthetic animals. Interhemispheric comparisons show clearly lower activity in the affected cortex of the amputated rat (lower intensity of the color) and similar in the prosthetic one. Also the volume of the neural aggregates dedicated to the process of information o fan individual whisker (show in sections and pointed by the arrows) has been reduced in amputee but maintained intact in the prosthetic

Our results are important in order to increase the neurobiological understanding of the plastic phenomena induced by peripheral input manipulation and to establish the scientific basis to properly model and predict all the complex phenomena that could be triggered by the implantation of the prosthesis and of different types of interfaces. Applications in the biomedical engineering field could range from definition of optimal stimulation parameters with measurable behavioural effects in future advanced neuroprostheses to the development of new neurorehabilitation therapies [2].

References

1. E. Ahissar, R. Sosnik, S. Haidarliu, Transformation from temporal to rate coding in a somatosensory thalamocortical pathway. Nature **406**(6793), 302–306 (2000)
2. S.L. Gonzalez Andino, C. Herrera-Rincon, F. Panetsos, R. Grave de Peralta, Combining BMI stimulation and mathematical modeling for acute stroke recovery and neural repair. Front. Neurosci. **5**(87), (2011)

3. B. Cavalieri, M. Ursinum, Geometria indivisibilibus continuorum nova quadam ratione promota. Kessinger Publishing (2010)
4. W.G. Cohran, *Sampling Techniques* (Wiley, NY, 1977)
5. L.M. Cruz-Orive, Precision of cavalieri sections and slices with local errors. J. Microsc. **193**(3), 182–198 (1999)
6. M.A. Delesse, Procédé mécanique pour déterminer la composition des roches. F. Savy (1866)
7. M.E. Diamond, M. Von Heimendahl, E. Arabzadeh, Whisker-mediated texture discrimination. PLoS Biol. **6**(8), e220 (2008)
8. M.S. Fee, P.P. Mitra, D. Kleinfeld, Automatic sorting of multiple unit neuronal signals in the presence of anisotropic and non-gaussian variability. J. Neurosci. Meth. **69**(2), 175–188 (1996)
9. M.S. Fee, P.P. Mitra, D. Kleinfeld, Variability of extracellular spike waveforms of cortical neurons. J. Neurophysiol. **76**(6), 3823 (1996)
10. G.L. Gerstein, M.J. Bloom, I.E. Espinosa, S. Evanczuk, M.R. Turner, Design of a laboratory for multineuron studies. IEEE Trans. Syst. Man Cybern. **13**(5), 668–676 (1983)
11. E.M. Glaser, Separation of neuronal activity by waveform analysis. Adv. Biomed. Eng. **1**, 77–136 (1971)
12. J.M. Goldberg, P.B. Brown, Response of binaural neurons of dog superior olivary complex to dichotic tonal stimuli: Some physiological mechanisms of sound localization. J. Neurophysiol. **32**(4), 613 (1969)
13. H.J. Gundersen, E.B. Jensen, The efficiency of systematic sampling in stereology and its prediction. J. Microsc. **147**(Pt 3), 229 (1987)
14. K.D. Harris, D.A. Henze, J. Csicsvari, H. Hirase, G. Buzsáki, Accuracy of tetrode spike separation as determined by simultaneous intracellular and extracellular measurements. J. Neurophysiol. **84**(1), 401 (2000)
15. C. Herrera-Rincon, C. Torets, A. Sanchez-Jimenez, C. Avendaño, P. Guillen, F. Panetsos, in *Structural Preservation of Deafferented Cortex Induced by Electrical Stimulation of a Sensory Peripheral Nerve*. Engineering in Medicine and Biology Society (EMBC), 2010. Annual International Conference of the IEEE (IEEE, NY, 2010), pp. 5066–5069
16. E. Hulata, R. Segev, E. Ben-Jacob, A method for spike sorting and detection based on wavelet packets and shannon's mutual information. J. Neurosci. Meth. **117**(1), 1–12 (2002)
17. Y. Karklin, M.S. Lewicki, A hierarchical bayesian model for learning nonlinear statistical regularities in nonstationary natural signals. Neural Comput. **17**(2), 397–423 (2005)
18. A. Lak, E. Arabzadeh, M.E. Diamond, Enhanced response of neurons in rat somatosensory cortex to stimuli containing temporal noise. Cerebr. Cortex **18**(5), 1085 (2008)
19. J.C. Letelier, P.P. Weber, Spike sorting based on discrete wavelet transform coefficients. J. Neurosci. Meth. **101**(2), 93–106 (2000)
20. M.S. Lewicki, A review of methods for spike sorting: the detection and classification of neural action potentials. Netw. Comput. Neural Syst. **9**(4), 53–78 (1998)
21. B.L. McNaughton, J. O'Keefe, C.A. Barnes, The stereotrode: A new technique for simultaneous isolation of several single units in the central nervous system from multiple unit records. J. Neurosci. Meth. **8**(4), 391–397 (1983)
22. P.R. Mouton, *Principles and Practices of Unbiased Stereology: An Introduction for Bioscientists* (Johns Hopkins University Press, MD, 2002)
23. F. Panetsos, A. Sanchez-Jimenez, Single unit oscillations in rat trigeminal nuclei and their control by the sensorimotor cortex. Neuroscience **169**(2), 893 – 905 (2010)
24. A. Pavlov, V.A. Makarov, I. Makarova, F. Panetsos, Sorting of neural spikes: When wavelet based methods outperform principal component analysis. Nat. Comput. **6**(3), 269–281 (2007)
25. G. Paxinos, C. Watson, *The Rat Brain in Stereotaxic Coordinates* (Academic, NY, 2007)
26. M.C. Quirk, M.A. Wilson, Interaction between spike waveform classification and temporal sequence detection. J. Neurosci. Meth. **94**(1), 41–52 (1999)
27. Q. Quiroga, Z. Nadasdy, Y. Ben-Shaul, Unsupervised spike detection and sorting with wavelets and superparamagnetic clustering. Neural Comput. **16**, 1661–1687 (2004)
28. R. Romo, A. Hernández, A. Zainos, C. Brody, E. Salinas, Exploring the cortical evidence of a sensory–discrimination process. Phil. Trans. Roy. Soc. Lond. B Biol. Sci. **357**(1424), 1039 (2002)

29. E. Salinas, A. Hernández, A. Zainos, R. Romo, Periodicity and firing rate as candidate neural codes for the frequency of vibrotactile stimuli. J. Neurosci. **20**(14), 5503 (2000)
30. A. Sanchez-Jimenez, F. Panetsos, A. Murciano, Early frequency-dependent information processing and cortical control in the whisker pathway of the rat: Electrophysiological study of brainstem nuclei principalis and interpolaris. Neuroscience **160**(1), 212–226 (2009)
31. E.M. Schmidt, Computer separation of multi-unit neuroelectric data: A review. J. Neurosci. Meth. **12**(2), 95–111 (1984)
32. R.K. Snider, A.B. Bonds, Classification of non-stationary neural signals. J. Neurosci. Meth. **84**(1–2), 155–166 (1998)
33. P.M.E. Waite, Chapter 26: Trigeminal Sensory System in The Rat Nervous System, 705–724, Elsevier Academic Press, third edition (2004)
34. M. Wong-Riley, Changes in the visual system of monocularly sutured or enucleated cats demonstrable with cytochrome oxidase histochemistry. Brain Res. **171**(1), 11–28 (1979)
35. M.T.T. Wong-Riley, Cytochrome oxidase: An endogenous metabolic marker for neuronal activity. Trends Neurosci. **12**(3), 94–101 (1989)
36. T.A. Woolsey, The structural organization of layer iv in the somatosensory region (si) of the mouse cerebral cortex: The description of a cortical field composed of discrete cytoarchitectonic units. Brain Res. **17**, 205–242 (1970)
37. T.A. Woolsey, C. Welker, R.H. Schwartz, Comparative anatomical studies of the sml face cortex with special reference to the occurrence of "barrels" in layer iv. J. Comp. Neurol. **164**(1), 79–94 (1975)
38. T.A. Woolsey, M.L. Dierker, D.F. Wann, Mouse smi cortex: Qualitative and quantitative classification of golgi-impregnated barrel neurons. Proc. Natl. Acad. Sci. U.S.A. **72**(6), 2165 (1975)

EEG Based Biomarker Identification Using Graph-Theoretic Concepts: Case Study in Alcoholism

Vangelis Sakkalis and Konstantinos Marias

Abstract Over the past few years there has been an increased interest in studying the under-lying neural mechanism of cognitive brain activity in order to identify features capable of discriminating brain engagement tasks in terms of cognitive load. Rather recently there is a growing belief that the noninvasive technique of high-resolution quantitative electroencephalography may provide features able to identify and quantify functional interdependencies among synchronized brain lobes based on graph-theoretic algorithms. In the emerging view of translational medicine, graph-theoretic measures and tools, currently used to describe large scale networks, can be potential candidates for future inclusion in a clinical trial setting. This paper discusses different families of graph theoretical measures able to capture the topology of brain networks as potential EEG-based biomarkers. As a case study, statistically significant graph-theoretic indices, capable of capturing and quantifying collective motifs in an alcoholism paradigm are identified and presented.

Mathematics Subject Classification (2010): Primary 54C40, 14E20; Secondary 46E25, 20C20

V. Sakkalis (✉)
Institute of Computer Science, Foundation for Research and Technology, Heraklion, Greece
e-mail: sakkalis@ics.forth.gr

K. Marias
Institute of Computer Science, Foundation for Research and Technology, Heraklion, Greece
e-mail: kmarias@ics.forth.gr

P.M. Pardalos et al. (eds.), *Optimization and Data Analysis in Biomedical Informatics*, 171
Fields Institute Communications 63, DOI 10.1007/978-1-4614-4133-5_9,
© Springer Science+Business Media New York 2012

1   Introduction

Understanding the spatiotemporal characteristics of brain activity has been the long-standing aim over the previous decades. Especially, during the past decade there has been a growing interest in studying and discovering specific prognostic and predictive biomarkers that tackle a wide variety of brain related pathologies and cognitive functions by means of both imaging and signal acquisition modalities. Functional Magnetic Resonance (fMRI) scanners, Near Infrared Optical Tomography (NIROT) and Electroencephalography (EEG), Magnetoencephalography (MEG) devices are capable of recording hemodynamic and neuronal signals, respectively. Imaging techniques achieve millimeter spatial resolution but suffer from rather high temporal resolution (more than a second). On the other hand, signal based techniques are able to record events in a millisecond timeframe, which is a strong advantage considering that an action potential takes approximately 0.5–130 ms to propagate across a single neuron. Hence, EEG and MEG are perfect candidates for extracting signal-based biomarkers targeting brain diseases that can be potentially applied in diagnosis and disease progression monitoring [41]. In addition, such biomarkers can act synergistically to molecular level biomarkers (genomics, proteomics and metabolomics) that are closely correlated to drug development.

Among the various frameworks proposed for efficient quantification and summarization of brain information flow [42], graph theoretical approaches offer a unique perspective of studying both local and distinct brain interactions [6, 48, 54]. Graph measures have been applied to topological analysis of brain functional networks and many of them reflect disease and statistically significant differences between healthy subjects and subjects with neuropathologies such as epilepsy, Alzheimer's disease, autism, Parkinson's disease, alcoholism, and schizophrenia [8, 40, 44, 45, 52, 59]. All these diseases have been associated with abnormal neural synchronization that systematically differs from those of healthy control subjects. Epilepsy has been associated with too high and too extended neural synchronization [21, 32, 51]. Patients with Alzheimer's disease show reduced synchronization [5,18,19,56]. Cognitive dysfunctions associated with autism are explained with reduced functional connectivity and neural synchronization [29, 39]. There are increasing amounts of data linking impaired motor processing in Parkinson's disease with excessive synchrony in basal ganglia-cortical loops [14, 55, 58]. Concerning schizophrenia, there is a growing body of evidence that the clinical symptoms and cognition dysfunctions observed in schizophrenia are caused by a disturbance in connectivity between different brain regions. In particular there is reduction in both local- and long-range synchronization [25, 44]. Additionally, a number of studies claim that there is a strong negative association between the characteristic path length (i.e., the average of shortest path lengths between each pair of vertices) of the resting-state brain functional network and the intelligence quotient (IQ), suggesting that human intellectual performance is likely to be related to how efficiently our brain integrates in-formation between multiple brain regions [24]. Such efficient networks that enable a rapid integration of information from local, specialized brain areas even when they are distant are characterized as small-world networks [53] in analogy to with the small world phenomenon initially witnessed in social systems [27]. Finally, brain tumors and especially low grade gliomas have been studied for possible disturbances using MEG [3]. Alcohol addiction and dependence is the exemplar case addressed in this study.

In summary, this work illustrates the potential of using graph theoretical approaches in analyzing and characterizing complex brain network topologies and reports the most significant graph theoretic measures, in terms of statistical power, capable of identifying the coupling differences of high resolution EEG channels in alcoholic and control subjects during a working memory task. Section 2 addresses the potential of using graph theoretical measures to study brain functional connectivity. Section 3 refers to some indicative techniques for measuring functional connectivity and defines representative families of graph measures. Application in an alcoholism case is presented in Section 4, while the final section concludes this chapter.

2   Why Use Graph Theoretical Markers?

2.1   Functional and Anatomical Connectivity Considerations

Most recent attempts to explain brain function focus on the functional interactions among the underlying distributed neural assemblies of different cerebral regions. More specifically, the activation of specialized brain neuronal populations and the coordinated activation of very large numbers of neurons within the distributed system of the cerebral cortex, commonly referred as the functional segregation and integration principle [10, 42] is central in cognitive neuroscience and connectivity analysis methodologies.

Mainly two different groups addressing brain connectivity can be defined. Those based on neuroanatomical structural landmarks and the ones attempting to detect and assess functional connectivity patterns. *Neuroanatomical connectivity* may be considered as fiber pathways tracking over extended regions of the brain that are in accordance with anatomical knowledge [20, 42]. Such pathways may be evaluated using neuroimaging techniques including Magnetic Resonance Imaging (MRI) based techniques commonly applied in Schizophrenia [2] and Alzheimer's disease [15]. Especially Diffusion Tensor Imaging (DTI), apart from structural and anatomical information, also enables fiber tracking of white matter tracts [11, 17]. Nevertheless, neuroanatomical connectivity is inherently difficult to define given the fact that at the microscopic scale of neurons, new synaptic connections or elimination of existing ones, are formed dynamically and are largely dependent on the function executed [34]. On the other hand, *functional connectivity* basically targets the temporal connection dependences aiming to unravel statistically significant dependences between distant brain regions [8, 42] and can be better evaluated using signal based techniques of high temporal resolution from widely used modalities, such as EEG and MEG. In contrast to anatomical connections, functional connections may evolve on a much quicker time scale and can reveal information on network organization underlying specific brain functions. For completeness, effective connectivity could also be mentioned, which is a rather new concept defined as the direct or indirect influence that one neural system exerts over another [16, 42]. In other words it describes the directed interactions between different brain regions according to a predefined model specifying the casual links.

Evidently, the long lasting focus of neuroscience community falls in the different methodologies for measuring functional connectivity [42] and searching for functional biomarkers [41] able to characterize such patterns. In this direction graph theory has been recently applied to neuroscience due to its ability to study brain network dynamics and structures by providing objective measures of the networks composed by functional links of different brain regions. However, the different methodologies available to extract the salient characteristics from a complex brain network topology remain a challenge in the graph theory community.

3   Methodological Aspects of EEG Graph Theoretic Biomarker Formation

3.1   Graph Definition

Graphs $G = (N, L)$ consist of a set of nodes N and a set of pairs of linked nodes called edges or links L. In brain modeling we assume that brain regions of interest may be considered as nodes and the corresponding interconnections as edges. The definition of nodes can be accustomed to the experimental design and may refer to the actual EEG electrodes or to independent sources and components that may be defined using source localization methods [12] or blind source separation techniques like ICA [45, 62]. The number of nodes in the network equals to the number of the sources or components under investigation. The estimation of edges is based on the calculation of a wide family of interdependence measures, henceforth referred as synchronization measures, that estimate the strength between all possible pairs of nodes. Such measures are often normalized and range between zero weight and one, with 0 meaning total independence/ absence of edge and 1 refer to maximum correlation/ strong interdependence. In this case we refer to *weighted graphs*. If strength information is discarded *binary graphs* are defined, where only the presence or absence of an edge is denoted. To switch to the binary case a threshold is applied, i.e., if edge weight is above threshold then the edge is kept, otherwise is removed. Threshold selection is not a straight forward task and there is no established and widely accepted way of favoring a specific threshold value. In practice, a broad range of threshold values is used to characterize the network. Automated ways of threshold selection are either based on statistics or based on signal techniques of

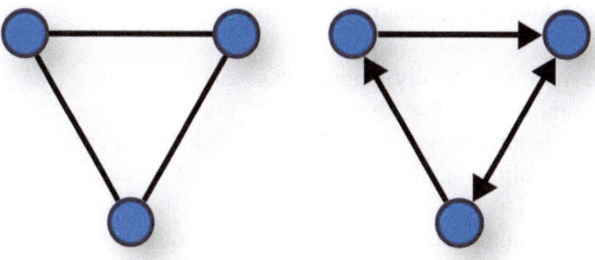

Fig. 1 A general example of an undirected (*left*) and a directed (*right*) graph with three nodes and three edges

selecting the optimal visualization threshold using surrogate (artificially generated ensembles of data aiming at revealing the most significantly coupled brain regions) datasets to correctly identify the most significant correlation patterns [45]. Finally, there is also a distinction between undirected and directed graphs. *Directed graphs* include ordered pairs of nodes, called arcs, directed edges, or arrows and reflect not only the presence of an edge but also whether there is coupling information available, i.e., one node drives another node (Fig. 1 right). *Undirected graphs* are defined in terms of unordered pairs of nodes known as edges (Fig. 1 left). Each graph is represented by its connectivity matrix that stores the weights between all possible pairs (i, j) of nodes and is referred as adjacency matrix. Depending on the type of the graph one may refer to either weighted (w_{ij}) or binary adjacency matrices. A summary of different ways for calculating the adjacency matrices using signal based functional connectivity estimation techniques follows next.

3.2 Brain Connectivity Analysis Using EEG

The application of high time resolution EEG may resolve interdependence patterns of cortical assemblies using linear interdependence measures, nonlinear synchronization estimators, and information-based techniques. The challenging problem of developing methods to efficiently and accurately quantify information processing mechanisms of the brain has been under study from the late 1960s [4]. In this section we only refer to some of the most widely used techniques along with some of their applications. In-depth details are provided in two such approaches (magnitude squared coherence and Phase Level Synchronization) that are further utilized in the experimental section in search of possible biomarkers for the alcoholism case.

3.2.1 Linear Coherence Estimation Technique

Coherence and more specifically magnitude squared coherence (MSC) is a linear measure that compared to the classical cross-correlation has the advantage of showing the covariation between two signals in distinct brain regions as a function of frequency; thus allowing the study of spatial correlations between different frequency bands [36].

Consider two simultaneously measured discrete time series x_n and y_n, $n = 1, \ldots, N$. The most commonly used linear synchronization method is the cross-correlation function (C_{xy}) defined as:

$$C_{xy}(\tau) = \frac{1}{N - \tau} \sum_{i=1}^{N-\tau} \left((x_i - \bar{x}) / \sigma_x \right) \left((y_{i+\tau} - \bar{y}) / \sigma_y \right) \tag{1}$$

where $\bar{x}$ and σ_x denote mean and variance, while τ is the time lag. MSC or simply coherence is the cross spectral density function S_{xy}, which is simply derived via the fast Fourier transform (FFT) of (1), normalized by their individual autospectral density functions. However, due to finite size of neural data one is forced to estimate the true spectrum, known as periodogram, using smoothing techniques (e.g., Welch's method). Thus, MSC is calculated as:

$$\gamma_{xy}(f) = \frac{|\langle S_{xy}(f) \rangle|^2}{|\langle S_{xx}(f) \rangle||\langle S_{yy}(f) \rangle|} \tag{2}$$

where $\langle \, \rangle$ indicates window averaging. The estimated MSC for a given frequency f ranges between 0 (no coupling) and 1 (maximum linear interdependence).

Other alternatives include wavelet based approaches i.e., Wavelet Coherence (WC) and more selective wavelet approaches that identify and extract only the statistically significant portion of the interacting signals as described in [43,44,49]. The WC method was successfully utilized to define biomarkers in the case of schizophrenia [44]. Other implementations include ARMA modeling approaches in estimating signal correlation by defining AR-coherence using a bivariate autoregressive process to describe the signals [47].

3.2.2 Nonlinear Synchronization and Information Based Approaches

A completely different approach in analyzing the nonlinear EEG dynamics emerged some decades after the discovery of deterministic chaos [26] when the notion of connectivity outreached synchronization phenomena of interacting nonlinear oscillators [35, 37]. Among the different available frameworks for capturing nonlinear synchronization, mainly the phase and generalized synchronization concepts are most widely used in several neuropathologies and cognitive research during the past years [46]. The most representative method capable of obtaining a statistical measure of the strength of phase synchronization is the Phase Locking Value (PLV) [22, 46]. The PLV approach assumes that two dynamic systems may have their phases synchronized even if their amplitudes are zero correlated [28]. The phase synchronization is defined as the locking of the phases associated to each signal, such as:

$$|\phi_x(t) - \phi_y(t)| = const.$$ (3)

In order to estimate the instantaneous phase of our signal, we transform it using the Hilbert transform (HT), whereby the analytical signal $H(t)$ is computed as:

$$H(t) = x(t) + i\tilde{x}(t)$$ (4)

where $\tilde{x}(t)$ is the HT of $x(t)$, defined as:

$$\tilde{x}(t)\frac{1}{\pi} PV \in_{-\infty}^{\infty} \frac{x(t')}{t-t'} dt'$$ (5)

where PV denotes the Cauchy principal value. The analytical signal phase is defined as:

$$\phi(t) = \arctan \frac{\tilde{x}(t)}{x(t)} .$$ (6)

Therefore for the two signals $x(t)$, $y(t)$ of equal time length with instantaneous phases $\phi_x(t), \phi_y(t)$ respectively the PLV bivariate metric is defined as:

$$PLV = \left| \frac{1}{N} \sum_{j=0}^{N-1} e^{i(\phi_x(j\Delta t) - \phi_y(j\Delta t))} \right|$$ (7)

where Δt is the sampling period and N is the sample number of each signal. PLV takes values within the $[0, 1]$ space, where 1 indicates perfect phase synchronization and 0 indicates lack of synchronization.

On the other hand *Generalized Synchronization* (GS) is based on the idea of measuring how neighborhoods (i.e., recurrences) in one chaotic attractor maps in-to the other. To form such attractors from the raw EEG data, we should first construct delay vectors using a procedure known as time-delay embedding [57]. Attractor mapping turned out to be the most robust and reliable way of assessing the extent of GS [1, 38, 46]. For an in-depth mathematical reasoning of the afore-mentioned techniques in an epileptic paradigm the interested reader is referenced in [46].

As a final point, we should also mention that wide use of a variety of information-based techniques that are susceptible to both linear and nonlinear signal interdependences. The underlying concept of these is that one is able to quantify the amount of information gained about one signal from measuring the other as a function of delay between these two signals. This is known as Cross Mutual Information (CMI) and has been successfully applied to Alzheimer's and Schizophrenia diseases [19, 30].

3.3   Graph Measures

Graph theoretical measures may be potentially considered as neurophysiological biomarkers because they can be extremely efficient for capturing and localizing brain activity motifs. They may be treated as features reflecting hidden signs capable of uniquely characterizing a disease or cognitive process. In this section we refer to three wide families of weighted graph theoretical measures that are informative about the brain network topology. The first two families are defined in analogy to the functional segregation and integration principle as discussed earlier (Sect. 2.1), whereas the third family assesses the importance of individual nodes in the brain network and is known as centrality measures. In our quest for identifying neurophysiological biomarkers for the case of the alcoholism we calculated and tested the following measures as candidate biomarkers. A recent overview of these measures can be found in [40]. It should be noted though that some of these measures are still limitedly used in the wider neuroscience research. Before proceeding to the exact definitions of these measures, let us first recall the notations used. A graph is defined as $G = (N, L)$, where N is the set of all the nodes, L is a set of pairs (i, j) of linked nodes $(i, j \in N)$ called edges. n is the number of nodes and l is the number of edges. W_{ij} denotes the normalized weighted adjacency matrix $0 \leq w_{ij} \leq 1$ as calculated from the functional connectivity techniques (Sect. 3.2) between all possible combinations of nodes. l^w is the sum of all weights in the network $l^w = \sum_{i,j \in N} w_{ij}$.

3.3.1   Integration Measures

The *shortest weighted path length* (distance) d_{ij}^w between two nodes i, j is the weighted sum of the minimum number of edges we need to traverse in order to go from node i to node j and is defined as:

$$d_{ij}^w = \sum_{a_{uv} \in g_{i \leftrightarrow j}^w} f(w_{uv}) \tag{8}$$

where $g_{i \leftrightarrow j}^w$ denotes the shortest weighted path between i, j (a_{uv} are the nodes we need to traverse) and f is a mapping function that transform weights to length estimations using for example an inverse function or $1 - \log_2(x)$. When no connection exists $d_{ij}^w = \infty$ for all disconnected nodes. This measure quantifies how easy it is for different brain regions to communicate. Thus it is strongly related to the paths that represent different routes of information flow between all possible nodes. Additional integration measures include the following:

Weighted characteristic path length [60]:

$$L^w = \frac{1}{n} \sum_{i \in N} \frac{\sum_{j \in N, j \neq i} d_{ij}^w}{n-1}. \tag{9}$$

It is the weighted average of edges in the shortest paths between every pair of nodes in the network.

Weighted global efficiency [23]:

$$E^w = \frac{1}{n} \sum_{i \in N} \frac{\sum_{j \in N, j \neq i} (d_{ij}^w)^{-1}}{n-1}. \tag{10}$$

3.3.2 Segregation Measures

The *weighted clustering coefficient* [33] represents the probability that neighbors of a node are also connected. In essence it reflects the tendency of the network to form local clusters and is defined as:

$$C^w = \frac{1}{n} \sum_{i \in N} \frac{2t_i^w}{k_i(k_i - 1)} \tag{11}$$

where $t_i^w = \frac{1}{2} \sum_{j,h \in N} (w_{ij}, w_{ih}, w_{jh})^{1/3}$ equals to the weighted geometric mean of triangles around i. k_i is the degree of vertex i and $k_i(k_i - 1)/2$ is the number of edges that could exist within its neighborhood. Schematically one may assess such a measure by calculating the fraction number of triangles (closed loops between neighboring nodes) around an individual node. Alternative definitions are found in [50]. Additional segregation measures include the following:

Weighted transitivity [31]:

$$T^w = \frac{\sum_{i \in N} 2t_i^w}{\sum_{i \in N} k_i(k_i - 1)}. \tag{12}$$

Weighted modularity [31]:

$$Q^w = \frac{1}{l^w} \sum_{i,j \in N} \left[w_{ij} - \frac{k_i^w k_j^w}{l^w} \right] \delta_{m_i, m_j} \tag{13}$$

where m_i is the module containing node i, and $\delta_{m_i, m_j} = 1$ if $m_i = m_j$, and 0 otherwise.

3.3.3 Centrality Measures

The *weighted degree* (k) is the most representative measure of centrality that quantifies how important, in terms of interaction frequency, a node role is for the whole brain network function. If a node functions as a hub and interacts frequently with many other nodes it is found to have a high degree. The weighted degree [60] denotes the sum of all weighted edges per node and is defined as:

$$k_i^w = \sum_{j \in N} w_{ij} \, . \tag{14}$$

Additional centrality measures include the following:

Weighted betweenness centrality [9]:

$$b_i^w = \frac{1}{(n-1)(n-2)} \sum_{h,j \in N, h \neq i, i \neq j} \frac{d_{hj}(i)}{d_{hj}} \tag{15}$$

where d_{hj} is the weighted sum of shortest paths between nodes h and j, and $d_{hj}(i)$ is the weighted sum of shortest paths between h and j that pass through i.

Weighted within-module degree z-score [13]:

$$z_i^w = \frac{k_i^w(m_i) - \bar{k}^w(m_i)}{\sigma^{k^w(m_i)}} \tag{16}$$

where m_i is the module containing node i, $k_i^w(m_i)$ is the number of links between i and all other nodes in m_i, and $\bar{k}(m_i)$, $\sigma^{k^w(m_i)}$ are the respective mean and standard deviation of the m_i module degree distribution.

Weighted participation coefficient [13]:

$$y_i^w = 1 - \sum_{m \in M} \left(\frac{k_i^w(m)}{k_i^w} \right)^2 \tag{17}$$

where M is the set of modules (see modularity in Sect. 3.3.2) and $k_i^w(m)$ is the weighted sum of edges between i and all nodes in module m.

4 Application in an Alcoholism Case

Impaired cognitive functioning has repeatedly been reported in alcohol-dependent individuals [61]. The fact that alcoholics have cognitive deficits in performing complex coordinated tasks suggests some related differentiation in brain functional connectivity as expressed by synchronization between different neural assemblies [7]. Hence, alcohol dependence is a treatable disease that could be studied using graph measures.

4.1 Experimental Setting

The EEG signals used in this work arise from 38 right-handed (control and alcoholic) subjects (had no personal or family history of any neurological disease, no age difference and normal vision or corrected normal vision) that were recorded in an electrically shielded, sound and light attenuated room. Participants were sitting in a reclined chair and fixated a point in the center of a computer display located 1m away from participants' eyes. Each subject was fitted with a 61-lead electrode cap (ECI, Electrocap International) according to the entire 10/20 International montage along with an additional 41 sites as follows: FPz, AFz, AF1, AF2, AFz, AF8, F1, F2, F5, F6, FCz, FC2, FC3, FC4, FC5, FC6, FC7, FC8, C1, C2, C5, C6, CPz, CP1, CP2, CP3, CP4, CP5, CP6, TP7, TP8, PI, P2, P5, P6, POz, PO1, PO2, PO7, and PO8 (Standard Electrode Position Nomenclature, American EEG Association 1990). All scalp electrodes were referred to Cz. Subjects were grounded with a nose electrode, and the electrode impedance was always below $5\,k\Omega$. Two additional bipolar deviations were used to record the vertical and horizontal EOG. The signals were amplified with a gain of 10,000 by Ep-A2 amplifiers (Sensorium, Inc., Charlotte, VT) with a bandpass between 0.02 and 50 Hz. The amplified signals were sampled at a rate of 256 Hz while each subject performs a picture rehearsal task as described below. Trials with excessive eye and body movements ($>73.3\,\mu V$) were rejected on-line and ten trials per subject where averaged together.

A working memory (WM) experiment was set up where each subject was exposed to pictures of objects chosen from the 1980 Snodgrass and Vanderwart picture set. These stimuli were randomized (but not repeated) and presented on a white background at the center of a computer monitor and were approximately 5–10 cm × 5–10 cm, thus subtending a visual angle of 0.05–0.1°. Ten trials were performed. The interval between each trial was fixed to 3.2 s. The participants were instructed to memorize the pictures in order to be able to identify them later.

4.2 Methodology and Results

Synchronous oscillations of certain types of such assemblies in different frequency bands were captured in this study using both linear (MSC) and nonlinear methods (PLV) to capture possible functional correlations and form the adjacency matrix. Then the above mentioned graph measures were successfully calculated and tested as indices of cerebral engagement in the alcoholism case. Other cognitive tasks or brain pathologies can be treated accordingly.

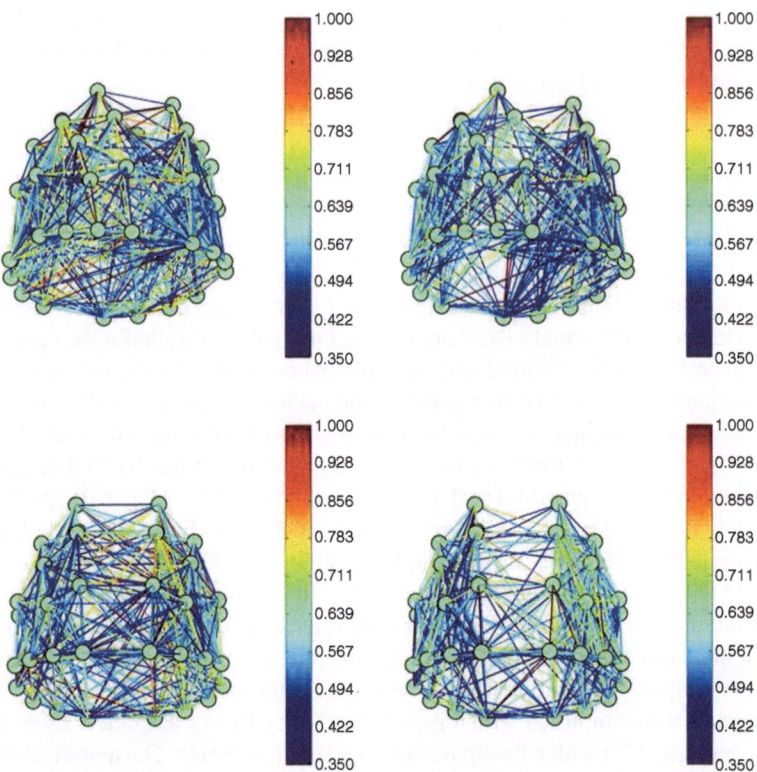

Fig. 2 A "healthy" average network topology (*top left*) appear to be much more dense than the alcoholic case (*top right*). *Bottom* images present a different 3D perspective of the topology. These graphs reflect broadband signal interdependence calculated using PLV

More specifically, for each population we accumulated separately the adjacency matrices of the experimental subjects, i.e., 38 controls and alcoholics in the WM task. For each subject in each group under consideration we calculated all nine graph measures (Lw, Ew, Cw, Tw, Qw, Kw, bw, zw, yw) that were treated as features. When measuring functional connectivity in neuroscience applications we often calculate connectivity in pre-filtered brain signals according to frequency bands of interest. These bands are typically defined as delta (0–4 Hz), theta (4–8 Hz), alpha (8–13 Hz), beta (13–30 Hz), lower gamma (30–45 Hz) and higher gamma (45–90 Hz). Frequency bands in neuroscience research are of paramount importance since they reflect rhythmic activity related to certain biological significance. In our application, apart from broadband unfiltered signals, we analyzed also alpha and beta bands, since related literature support most prominent findings in alpha and beta frequency bands [7, 45].

A visualization of the averaged adjacency matrices and 3D graph topology in both "healthy" and "alcoholic" brain using the PLV method applied in broadband signals (no filtering) and beta band signals is illustrated in Figs. 2 and 3, respectively.

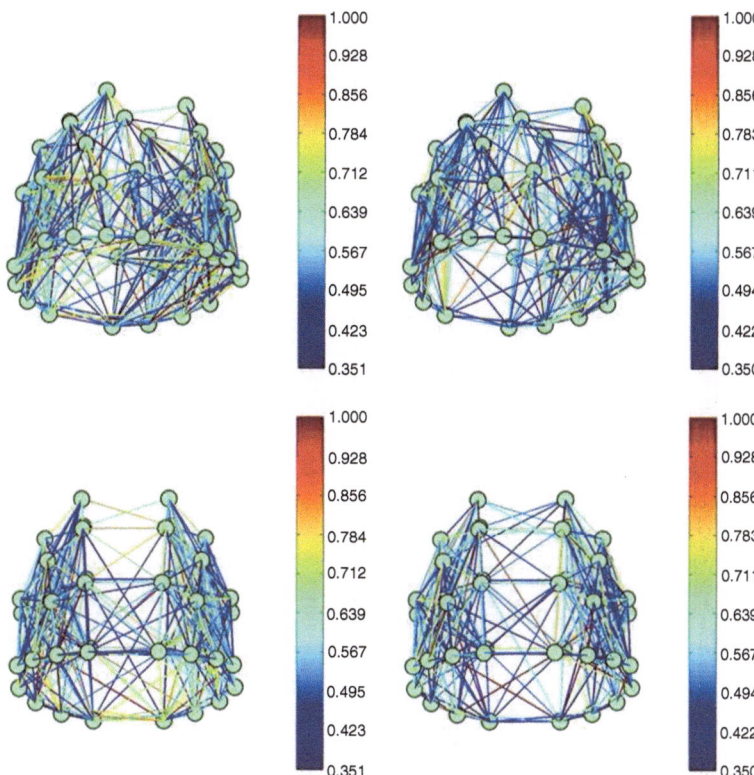

Fig. 3 When beta band signal interdependence is calculated using PLV the connections are less pronounced than the broadband case. Here again "healthy" average network topologies appear on the *left column* and the alcoholic counterpart on the *right*

It can be easily illustrated using such a framework that an alcoholic subject exhibits impaired synchronization of brain activity and loss of direct connections during the rehearsal process as compared to a healthy one (Fig. 2). Lower synchronization was also evident in the alcoholic subject, most prominently in beta band (Fig. 3), as compared to a control. Here again loss of connections is evident in the alcoholic case. Similar findings were reported also in de Bruin et al. [7].

Hence, we end up with 39x9 features for each group and frequency band. To further identify which of these possible features may also be considered as a potential biomarker we performed statistical tests for significance under the null hypothesis that there is no significant difference between the population mean (when normality is met) and median (when the features are not normally distributed) of the groups. Normality of the distributions was tested using the D' Agostino-Pearson's test. When normality was met, we performed t-tests to enhance the statistical power of our results. Otherwise, when normality was not met, we performed Mann-Whitney U-tests. Table 1 summarizes the outcome of the runs. For the cases where null hypothesis was rejected we also state the calculated p-value.

Table 1 Summary of the statistical significance achieved using various graph theoretical measures ((9)–(17)) applied in different methods for estimating functional connectivity (MSC, PLV). No filtering was performed in the EEG signals. p-values are reported when statistical significance was achieved ($p < 0.05$) under the corresponding test. t-test (TT) was used when normality was met and Mann-Whitney U-test (UT) when no normal sample distribution was found. X denotes no statistical significance

	Integration measures		Segregation measures				Centrality measures		
	L^w	E^w	C^w	T^w	Q^w	K^w	b^w	z^w	y^w
MSC	UT $p < 0.001$	UT $p < 0.0001$	UT $p < 0.0001$	UT $p < 0.0001$	UT $p < 0.0001$	UT $p < 0.0001$	UT $p < 0.0001$	X–UT $p < 0.0001$	UT $p < 0.001$
PLV	UT $p < 0.026$	TT $p < 0.026$	TT $p < 0.016$	TT $p < 0.019$	X–TT	TT $p < 0.014$	X–UT	X–TT	UT $p < 0.008$

Table 2 Summary of the statistical significance achieved using various graph theoretical measures ((9)–(17)) applied in different methods of estimating functional connectivity (MSC, PLV). Alpha band EEG signals are considered. p-values are reported when statistical significance was achieved ($p < 0.05$) under the corresponding test. t-test (TT) was used when normality was met and Mann-Whitney U-test (UT) when no normal sample distribution was found. X denotes no statistical significance

	Integration measures			Segregation measures			Centrality measures		
	L^w	E^w	C^w	T^w	Q^w	K^w	b^w	z^w	y^w
MSC	X–TT	X–UT	X–UT	X–UT	X–TT	X–UT	X–TT	X–TT	X–UT
PLV	X–TT	UT	UT	UT	X–TT	UT	X–UT	X–TT	X–TT
		$p<0.033$	$p<0.032$	$p<0.032$		$p<0.032$			

Table 3 Summary of the statistical significance achieved using various graph theoretical measures ((9)–(17)) applied in different methods of estimating functional connectivity (MSC, PLV). Beta band EEG signals are considered. p-values are reported when statistical significance was achieved ($p < 0.05$) under the corresponding test. t-test (TT) was used when normality was met and Mann-Whitney U-test (UT) when no normal sample distribution was found. X denotes no statistical significance

	Integration measures			Segregation measures			Centrality measures		
	L^w	E^w	C^w	T^w	Q^w	K^w	b^w	z^w	y^w
MSC	TT	TT	UT	UT	X–UT	TT	X–UT	X–TT	UT
	$p<0.008$	$p<0.023$	$p<0.008$	$p<0.008$		$p<0.029$			$p<0.026$
PLV	TT	UT	UT	UT	X–UT	UT	X–TT	X–TT	UT
	$p<0.002$	$p<0.004$	$p<0.002$	$p<0.002$		$p<0.002$			$p<0.002$

Some initial conclusions drawn from the Tables 1, 2 and 3 indicate that several graph theoretical measures are statistically significant, meaning that the underlying brain networks as captured from these measures are different in healthy controls than in alcoholics. The proposed synchronization analysis in combination with the network analysis and visualization are able to picture with increased certainty, the brain network topology during a certain mental task. Both linear (MSC) and nonlinear (PLV) interdependence patterns were able to capture differentiations. Broadband signals and beta filtered signals revealed significant differences when integration measures (L^w, E^w) segregation measures (C^w, T^w) and centrality measures (K^w, y^w) were estimated. More specifically, MSC was more successful with integration measures in beta band, whereas PLV was better in broadband signals (t-test was able to discriminate the two populations with increased statistical power as opposed to the limited statistical power of Mann-Whitney U-test). Alpha band was not able to identify statistically significant differentiations using MSC. PLV was able to identify only some significant measures. Also in beta band PLV was able to identify stronger differentiations (in terms of statistical power) with smaller p-values than MSC. Similar findings were reported during moderate-to-heavy-alcohol intake during rest and mental rehearsal in the recent work of de Bruin et al. [58].

As a conclusion, our experiments indicated that segregation and centrality measures are getting lower in the case of the alcoholics as compared to healthy controls suggesting that an alcoholic brain has significantly less node (EEG channel) connections while performing the same mental task as the control one. This results in a more efficient brain network in the case of healthy controls as opposed to alcoholics. The latter is also validated by the global efficiency measure that is lower in alcoholics than in controls.

5 Conclusions

In this work we investigated the use of EEG signal synchronization studies in combination with graph-theoretic approaches devised to study and stress the coupling dynamics of task-performing brain dynamical networks. The presented methodology was based on statistical significance testing in an alcoholism case study during mental rehearsal of pictures, which is known to reflect synchronization impairment, using both linear and nonlinear interdependence estimation measures. Graph statistical parameters were successful in capturing and quantifying collective but differentiated motifs present in the functional brain network of both healthy controls and alcoholics. Our analysis did not aim to study in depth the addictive disorder of alcohol dependence, but to demonstrate the benefits of the possibility of using graph indices as possible biomarkers. Additional studies, involving larger data sets, are needed to explore and validate these findings further. Undoubtedly, the use of such measures as potential biomarkers could have a significant impact on the field, but currently there are considerable risks related to their premature adoption in clinical practice. Although significant steps towards the discovery of EEG neurophysiologically inspired biomarkers have been made, the ideal paradigms still need to be determined. However, basic neuroscience and brain knowledge discovery fields can certainly benefit from becoming familiar with the most robust methods for constructing statistically significant functional networks, as well as with measures for comparing and characterizing brain networks that may be proved to be valuable biomarkers of brain pathophysiology.

Acknowledgements The authors would like to thank Henri Begleiter at the Neurodynamics Laboratory, State University of NY Health Center at Brooklyn for kindly providing the EEG dataset.

References

1. J. Arnhold, P. Grassberger, K. Lehnertz, C.E. Elger, A robust method for detecting interde-pendences: application to intracranially recorded EEG. Phys. D: Nonlinear Phenom. **134**(4), 419–430 (1999)

2. D.S. Bassett, E. Bullmore, B.A. Verchinski, V.S. Mattay, D.R. Weinberger, A. Meyer-Lindenberg, Hierarchical organization of human cortical networks in health and schizophrenia. J. Neurosci. **28**(37), 9239 (2008)
3. I. Bosma, J.C. Reijneveld, M. Klein, L. Douw, B.W. Van Dijk, J.J. Heimans, C.J. Stam, Disturbed functional brain networks and neurocognitive function in low-grade glioma patients: A graph theoretical analysis of resting-state MEG. Nonlinear Biomed. Phys. **3**(1), 9 (2009). doi: 10.1186/1753-4631-3-9
4. M.A.B. Brazier, J.U. Casby, Crosscorrelation and autocorrelation studies of electroencephalographic potentials. Electroencephalogr. Clin. Neurophysiol. **4**(2), 201–211 (1952)
5. R.L. Buckner, J. Sepulcre, T. Talukdar, F.M. Krienen, H. Liu, T. Hedden, J.R. Andrews-Hanna, R.A. Sperling, K.A. Johnson, Cortical hubs revealed by intrinsic functional connectivity: Mapping, assessment of stability, and relation to Alzheimer's disease. J. Neurosci. **29**(6), 1860 (2009)
6. E. Bullmore, O. Sporns, Complex brain networks: Graph theoretical analysis of structural and functional systems. Nat. Rev. Neurosci. **10**(3), 186–198 (2009)
7. E.A. de Bruin, C.J. Stam, S. Bijl, M.N. Verbaten, J.L. Kenemans, Moderate-to-heavy alcohol intake is associated with differences in synchronization of brain activity during rest and mental rehearsal. Int. J. Psychophysiol. **60**(3), 304–314 (2006)
8. A.A. Fingelkurts, A.A. Fingelkurts, S. Kahkonen, New perspectives in pharmaco-electroencephalography. Progr. Neuro Psychopharmacol. Biol. Psychiatr. **29**(2), 193–199 (2005)
9. L.C. Freeman, Centrality in social networks conceptual clarification. Soc. Network. **1**(3), 215–239 (1979)
10. K.J. Friston, Modalities, modes, and models in functional neuroimaging. Science **326**(5951), 399 (2009)
11. G. Gong, Y. He, L. Concha, C. Lebel, D.W. Gross, A.C. Evans, C. Beaulieu, Mapping anatomical connectivity patterns of human cerebral cortex using in vivo diffusion tensor imaging tractography. Cerebr. Cortex **19**(3), 524 (2009)
12. R. Grech, T. Cassar, J. Muscat, K.P. Camilleri, S.G. Fabri, M. Zervakis, P. Xanthopoulos, V. Sakkalis, B. Vanrumste, Review on solving the inverse problem in EEG source analysis. J. NeuroEng. Rehabilitation **5**(1), 25 (2008)
13. R. Guimera, L.A.N. Amaral, Cartography of complex networks: Modules and universal roles. J. Stat. Mech. Theor. Exp. **2005**, P02001 (2005)
14. C. Hammond, H. Bergman, P. Brown, Pathological synchronization in Parkinson's disease: Networks, models and treatments. Trends Neurosci. **30**(7), 357–364 (2007)
15. Y. He, Z. Chen, A. Evans, Structural insights into aberrant topological patterns of large-scale cortical networks in Alzheimer's disease. J. Neurosci. **28**(18), 4756 (2008)
16. B. Horwitz, The elusive concept of brain connectivity. Neuroimage **19**(2), 466–470 (2003)
17. Y. Iturria-Medina, EJ Canales-Rodriguez, L. Melie-Garcia, PA Valdes-Hernandez, E. Martinez-Montes, Y. Aleman-Gomez, JM Sanchez-Bornot, Characterizing brain anatomical connections using diffusion weighted MRI and graph theory. Neuroimage **36**(3), 645–660 (2007)
18. J. Jeong, EEG dynamics in patients with Alzheimer's disease. Clin. Neurophysiol. **115**(7), 1490–1505 (2004)
19. J. Jeong, J.C. Gore, B.S. Peterson, Mutual information analysis of the EEG in patients with Alzheimer's disease. Clin. Neurophysiol. **112**(5), 827–835 (2001)
20. M.A. Koch, D.G. Norris, M. Hund-Georgiadis, An investigation of functional and anatomical connectivity using magnetic resonance imaging. NeuroImage **16**(1), 241–250 (2002)
21. M.A. Kramer, E.D. Kolaczyk, H.E. Kirsch, Emergent network topology at seizure onset in humans. Epilepsy Res. **79**(2–3), 173–186 (2008)
22. J.P. Lachaux, E. Rodriguez, J. Martinerie, F.J. Varela, Measuring phase synchrony in brain signals. Hum. Brain Mapp. **8**(4), 194–208 (1999)
23. V. Latora, M. Marchiori, Efficient behavior of small-world networks. Phys. Rev. Lett. **87**(19), 198701 (2001)

24. Y. Li, Y. Liu, J. Li, W. Qin, K. Li, C. Yu, T. Jiang, Brain anatomical network and intelligence. PLoS Comput. Biol. **5**(5), e1000395 (2009)
25. Y. Liu, M. Liang, Y. Zhou, Y. He, Y. Hao, M. Song, C. Yu, H. Liu, Z. Liu, T. Jiang, Disrupted small-world networks in schizophrenia. Brain **131**(4), 945 (2008)
26. E.N. Lorenz, Deterministic nonperiodic flow. J. Atmos. Sci. **20**(2), 130–141 (1963). doi: http://dx.doi.org/10.1175/1520-0469(1963)020<0130:DNF>2.0.CO;2
27. S. Milgram, The small world problem. Psychol. Today **2**(1), 60–67 (1967)
28. F. Mormann, K. Lehnertz, P. David, Mean phase coherence as a measure for phase synchronization and its application to the EEG of epilepsy patients. Phys. D: Nonlinear Phenom. **144**(3–4), 358–369 (2000)
29. M. Murias, S.J. Webb, J. Greenson, G. Dawson, Resting state cortical connectivity reflected in EEG coherence in individuals with autism. Biol. Psychiatr. **62**(3), 270–273 (2007)
30. S.H. Na, S.H. Jin, S.Y. Kim, B.J. Ham, EEG in schizophrenic patients: mutual information analysis. Clin. Neurophysiol. **113**(12), 1954–1960 (2002)
31. M.E.J. Newman, Modularity and community structure in networks. Proc. Natl. Acad. Sci. **103**(23), 8577 (2006)
32. E. Niedermeyer, F.H.L. Da Silva, *Electroencephalography: Basic Principles, Clinical Applications, and Related Fields* (Lippincott Williams & Wilkins, PA, 2005)
33. J.P. Onnela, J. Saramäki, J. Kertész, K. Kaski, Intensity and coherence of motifs in weighted complex networks. Phys. Rev. E **71**(6), 065103 (2005)
34. A. Ooyen, Competition in the development of nerve connections: A review of models. Netw. Comput. Neural Syst. **12**(1), 1–47 (2001)
35. L.M. Pecora, T.L. Carroll, Synchronization in chaotic systems. Phys. Rev. Lett. **64**(8), 821–824 (1990)
36. G. Pfurtscheller, C. Andrew, Event-related changes of band power and coherence: methodology and interpretation. J. Clin. Neurophysiol. **16**(6), 512 (1999)
37. A.S. Pikovsky, On the interaction of strange attractors. Zeitschrift für Phys. B Condens. Matter **55**(2), 149–154 (1984)
38. R.Q. Quiroga, A. Kraskov, T. Kreuz, P. Grassberger, Performance of different synchronization measures in real data: A case study on electroencephalographic signals. Phys. Rev. E **65**(4), 041903 (2002)
39. G. Rippon, J. Brock, C. Brown, J. Boucher, Disordered connectivity in the autistic brain: Challenges for the new psychophysiology. Int. J. Psychophysiol. **63**(2), 164–172 (2007)
40. M. Rubinov, O. Sporns, Complex network measures of brain connectivity: uses and interpretations. Neuroimage **52**(3), 1059–1069 (2010)
41. V. Sakkalis, Applied strategies towards EEG/MEG biomarker identification in clinical and cognitive research. Biomarkers Med., Future Med. **5**(1), 93–105 (2011)
42. V. Sakkalis, "Review of Advanced Techniques for the estimation of Brain Connectivity measured with EEG/MEG", Comput Biol Med. **41**(12), 1110–1117 (2011). DOI:10.1016/j.compbiomed.2011.06.020
43. V. Sakkalis, M. Zervakis, S. Micheloyannis, Significant EEG features involved in mathematical reasoning: Evidence from wavelet analysis. Brain Topogr. **19**(1), 53–60 (2006)
44. V. Sakkalis, T. Oikonomou, E. Pachou, I. Tollis, S. Micheloyannis, M. Zervakis, in *Time-Significant Wavelet Coherence for the Evaluation of Schizophrenic Brain Activity Using a Graph Theory Approach*. Engineering in Medicine and Biology Society, 2006. EMBS'06. 28th Annual International Conference of the IEEE (IEEE, NY, 2006), pp. 4265–4268
45. V. Sakkalis, V. Tsiaras, M. Zervakis, I. Tollis, in *Optimal Brain Network Synchrony Visualization: Application in an Alcoholism Paradigm*. Engineering in Medicine and Biology Society, 2007. EMBS 2007. 29th Annual International Conference of the IEEE (IEEE, NY, 2007), pp. 4285–4288
46. V. Sakkalis, C.D. Giurcaneanu, P. Xanthopoulos, M.E. Zervakis, V. Tsiaras, Y. Yang, E. Karakonstantaki, S. Micheloyannis, Assessment of linear and nonlinear synchronization measures for analyzing EEG in a mild epileptic paradigm. IEEE Trans. Inform. Tech. Biomed. **13**(4), 433–441 (2009)

47. V. Sakkalis, T. Cassar, M. Zervakis, C.D. Giurcaneanu, C. Bigan, S. Micheloyannis, K.P. Camilleri, S.G. Fabri, E. Karakonstantaki, K. Michalopoulos, A decision support framework for the discrimination of children with controlled epilepsy based on EEG analysis. J. NeuroEng. Rehabilitation **7**, 24 (2010)
48. V. Sakkalis, V. Tsiaras, I. Tollis, Graph analysis and visualization for brain function characterization using EEG data. J. Healthc. Eng. **1**(3), 435-460 (2010)
49. V. Sakkalis, M. Zervakis, in *Methodological Framework for EEG Feature Selection Based on Spectral and Temporal Profiles*, ed. by W. Chaovalitwongse, P.M. Pardalos, P. Xanthopoulos. Computational Neuroscience (Springer, New York, 2010), pp. 43–56
50. J. Saramäki, M. Kivelä, J.P. Onnela, K. Kaski, J. Kertész, Generalizations of the clustering coefficient to weighted complex networks. Phys. Rev. E **75**(2), 027105 (2007)
51. K.A. Schindler, S. Bialonski, M.T. Horstmann, C.E. Elger, K. Lehnertz, Evolving functional network properties and synchronizability during human epileptic seizures. Chaos: An Interdiscipl. J. Nonlinear Sci. **18**, 033119 (2008)
52. W.W. Seeley, R.K. Crawford, J. Zhou, B.L. Miller, M.D. Greicius, Neurodegenerative diseases target large-scale human brain networks. Neuron **62**(1), 42–52 (2009)
53. O. Sporns, J.D. Zwi, The small world of the cerebral cortex. Neuroinformatics **2**(2), 145–162 (2004)
54. O. Sporns, G. Tononi, G.M. Edelman, Theoretical neuroanatomy: Relating anatomical and functional connectivity in graphs and cortical connection matrices. Cerebr. Cortex **10**(2), 127 (2000)
55. D. Stoffers, J.L.W. Bosboom, J.B. Deijen, E.C. Wolters, C.J. Stam, H.W. Berendse, Increased cortico-cortical functional connectivity in early-stage Parkinson's disease: An MEG study. Neuroimage **41**(2), 212–222 (2008)
56. K. Supekar, V. Menon, D. Rubin, M. Musen, M.D. Greicius, Network analysis of intrinsic functional brain connectivity in Alzheimers disease. PLoS Comput. Biol. **4**(6), e1000100 (2008)
57. F. Takens, Detecting strange attractors in turbulence. Dyn. Syst. Turbulence Warwick **1980**, 366–381 (1981)
58. L. Timmermann, E. Florin, C. Reck, Pathological cerebral oscillatory activity in parkinsons disease: A critical review on methods, data and hypotheses. Expert Rev. Med. Dev. **4**(5), 651–661 (2007)
59. P.J. Uhlhaas, C. Haenschel, D. Nikolić, W. Singer, The role of oscillations and synchrony in cortical networks and their putative relevance for the pathophysiology of schizophrenia. Schizophr. Bull. **34**(5), 927 (2008)
60. D.J. Watts, S.H. Strogatz, Collective dynamics of small-worldnetworks. Nature **393**(6684), 440–442 (1998)
61. G. Winterer, M.A. Enoch, KV White, M. Saylan, R. Coppola, D. Goldman, EEG phenotype in alcoholism: Increased coherence in the depressive subtype. Acta Psychiatr. Scand. **108**(1), 51–60 (2003)
62. M. Zervakis, K. Michalopoulos, V. Iordanidou, V. Sakkalis, Intertrial coherence and causal interaction among independent EEG components. J. Neurosci. Med. **197**(2), 302–314 (2011)

Maximal Connectivity and Constraints in the Human Brain

Roman V. Belavkin

Abstract We represent neural networks by directed graphs and consider the problem of maximal connectivity with constraints. This problem is motivated by some conflicting objectives in the design of biological neural networks. Inequalities and equations derived are tested on data and numerical estimates for parameters of a human brain. Results support an intuition that human brain is maximally connected subject to constraints on in- and out-degrees.

Mathematics Subject Classification (2010): Primary 94C15; Secondary 92C20

1 Introduction

The idea that graph-theoretic concepts can be used to analyze brain networks is not new. There has been a lot of research and interest in applying graph-theoretic methods in neuroscience (e.g., see [3] for review). This work considers brain network as a solution to an optimization problem with conflicting objectives. On one hand, there is an objective to represent and communicate information about the environment with the highest possible quality in order to achieve optimal or nearly optimal control of the body. This objective leads to maximization of connectivity between information sources (sensors) and information sinks (controls). On the other hand, there is an objective to minimize material and energy consumption of the brain. This objective leads to minimization of connectivity between neurons and their number. Thus, our hypothesis is that brain is designed to achieve a certain equilibrium in connectivity.

R.V. Belavkin (✉)
School of Engineering and Information Sciences, Middlesex University, London NW4 4BT, UK
e-mail: R.Belavkin@mdx.ac.uk

P.M. Pardalos et al. (eds.), *Optimization and Data Analysis in Biomedical Informatics*,
Fields Institute Communications 63, DOI 10.1007/978-1-4614-4133-5_10,
© Springer Science+Business Media New York 2012

In this work, we shall use this idea of extreme connectivity to derive some simple relations between parameters of a network. In particular, these will be relations between the numbers of inputs and outputs of a network and the numbers of hidden nodes and their degrees. In addition, we shall test the hypothesis and relations by predicting the corresponding parameters of a human nervous system.

The next section introduces the notation and reminds some basic concepts about directed graphs. The main results are presented in Sect. 3, some of which are new and some clarify derivations of formulae, presented earlier in [2]. The relations derived will be evaluated in the examples using numerical estimates of parameters in human nervous system from [2], which are overviewed in the Appendix. In conclusion, we shall discuss the results and some future directions of the work.

2 Notation

We consider a neural network as a directed graph $G = (V, E)$, where V is the set of vertexes or nodes, and $E \subset V \times V$ is the set of edges or arrows. This is because signal transmission in neurons with chemical synapses occurs in one direction. Each neuron is represented by a node $v \in V$, and each connection (axon) from v_i to v_j is represented by an arrow $e = (v_i, v_j) \in E$. We denote by $\deg^+(v)$ and $\deg^-(v)$ the in- and out-degrees of node $v \in V$ respectively. Recall the fundamental formula for the number of arrows in a directed graph:

$$|E| = \sum_{v \in V} \deg^+(v) = \sum_{v \in V} \deg^-(v). \tag{1}$$

Sources are nodes with $\deg^+(v) = 0$, and they represent inputs into the network. Sinks are nodes with $\deg^-(v) = 0$, and they represent outputs from the network. Other nodes with $\deg^+(v) > 0$ and $\deg^-(v) > 0$ represent the so-called hidden nodes. Let us denote the number of input, output and hidden nodes by m, n and s respectively:

$$m := |\{v \in V : \deg^+(v) = 0\}|$$
$$n := |\{v \in V : \deg^-(v) = 0\}|$$
$$s := |\{v \in V : \deg^+(v) > 0 \text{ and } \deg^-(v) > 0\}|.$$

Clearly, $|V| = m + n + s$.

Our aim in this paper is first to derive some relations between parameters m, n and s based on some assumptions about the connectivity and related constraints. Then we shall verify if these relations can predict some numerical parameters of the human brain based on biological data.

3 Maximizing Connectivity Under Constraints

3.1 Optimal Coding and Communication of Input States

A neural network implements a (possibly non-linear) transformation of an input signal $x \in X$ into an output $y \in Y$. This transformation can be deterministic, described by a function $y = f(x)$, or stochastic, described by conditional probability $P(y \mid x)$. Thus, a network can be considered as a communication channel. An optimal transformation maximizes some utility function. Here we shall assume that the utility is mutual information between x and y:

$$I(x, y) := \sum_{X \times Y} \left[\ln \frac{P(x, y)}{P(x)P(y)} \right] P(x, y).$$

It is well-known that without additional constraints, maximum (or supremum) of information is communicated by an injective function [5]; a non-injective function is in some sense equivalent to a noisy channel. The image of an injective function has the same cardinality as its domain: $|f(X)| = |X|$. It is not difficult to see that a neural network with $m > n$ cannot be an injective function. Indeed, if $\{1, \ldots, \alpha\}$ is the alphabet of α symbols that can be communicated over each edge, then α^m is the cardinality of the input into a network from m sources, and α^n is the cardinality of the output into n sinks.

However, if the input variable x is not completely random, so that its entropy $H\{x\} := - \sum [\ln P(x)] \, P(x) < \ln |X|$, then it is possible to communicate information perfectly by a noisy channel $P(y \mid x)$ [5, 6]. The output of such a channel must have cardinality no less than $e^{H\{x\}} < |X|$. Thus, a network with $n \geq H\{x\}/\ln \alpha$ output nodes can in principle communicate without loss of information.

To implement optimal transformation $P(y \mid x)$, one has to know or learn the distribution $P(x)$. Therefore, we conjecture that the network must be able to encode all input states x by its internal states. This suggests the following relation between sources and hidden nodes.

Proposition 1. *Let $G = (V, E)$ be a directed graph with m sources, and let $\deg^-(v) = l$ on average for all sources and hidden nodes. Then hidden nodes of G can represent all input information if their number is $s \geq ml$.*

Proof. Let α be the number of symbols communicated over each arrow in G. Then α^m is the number of input states (sentences) to be encoded. Each source and hidden node sends the same symbol α to $\deg^-(v) = l$ nodes on average, and $\deg^-(v) = l$ hidden nodes on average receive the same symbol. Thus, s hidden nodes can encode on average $\alpha^{s/l}$ states (sentences), and the result is obtained from inequality $\alpha^{s/l} \geq \alpha^m$.

Remark 2. Biological neurons can communicate only one symbol at a time, because they have only one axon. The axon, however, can branch and connect to many other neurons. Thus, node v with $\deg^+(v) = k$ inputs has the capacity to receive up to α^k different sentences, but can transmit further only α on its output. If there are l nodes connected to the same k inputs, then α^l messages can be communicated. This in turn requires that k nodes have $\deg^-(v) = l$.

Example 3. For $s \approx 10^{11}$ and $m \approx 4.84 \cdot 10^7$ of human nervous system (see Appendix), Proposition 1 gives an estimate $l \leq s/m \approx 2 \cdot 10^4$. This suggests that neurons in human CNS have similar values of $\deg^-(v)$ and $\deg^+(v)$. On the other hand, setting $l = 2 \cdot 10^3$ gives $s \geq .97 \cdot 10^{11}$.

3.2   Maximal Connectivity with Constraints

Maximally connected graph has the maximum number of edges, and each node has the maximal degree. In a biological network, maximal connectivity would require maximal amount of material and perhaps energy consumption. Let us consider a graph maximizing connectivity subject to constraints. The following relation can be obtained.

Proposition 4. *Let $G = (V, E)$ be a directed graph with m sources and n sinks. Then the number of hidden nodes in G with maximal connectivity subject to constraints $\deg^+(v) \leq k$ and $\deg^-(v) \leq l$ is*

$$s = \frac{nk - ml}{l - k}. \tag{2}$$

Proof. If G has connectivity such that $\deg^+(v) = k$ for all hidden nodes and sinks and $\deg^-(v) = l$ for all hidden nodes and sources, then (1) gives the following equality

$$|E| = (s + n)k = (s + m)l.$$

Thus, $nk - ml = (l - k)s$, which gives the desired result.

Example 5. Substituting numerical estimates for m and n in human CNS (see Appendix), and using $k = 2 \cdot 10^3$, $l = k - 1$ gives $s \approx 0.96 \cdot 10^{11}$, which is quite close to the estimated number of neurons in a human brain (10^{11}).

3.3   Length of an Input–Output Path

In this section, we analyze the length of a path from sources to sinks in a maximally connected graph with constraints on connectivity using a feed-forward network model. In such a network, nodes are grouped into layers $i = 0, 1, 2, \ldots, h, h + 1$. Layer $i = 0$ consists of m input nodes (sources), while layer $i = h + 1$ consists of n output nodes (sinks). Thus, h is the number of hidden layers, and the length of a path from sources to sinks is $h + 1$. In a feed-forward network, each node in layer i can only have connections with nodes in layer $i + 1$. Although this model ignores other connections (e.g., lateral, backward, forward connections), it allows us to derive a simple relation for the length of a path between sources and sinks.

Proposition 6. *Let $G = (V, E)$ be a feed-forward network with m sources and n sinks. Then the number of hidden layers in G with maximal connectivity subject to constraints $\deg^+(v) \le k$ and $\deg^-(v) \le l$ is*

$$h = \frac{\ln n - \ln m}{\ln l - \ln k} - 1. \tag{3}$$

The number of hidden nodes is

$$s = m \sum_{i=1}^{h} \left(\frac{l}{k}\right)^i = n \sum_{i=1}^{h} \left(\frac{l}{k}\right)^{-i}. \tag{4}$$

Proof. Let r_i denote the number of nodes in ith layer. The following relations hold

$$r_i \le k r_{i+1} \quad \text{and} \quad l r_i \ge r_{i+1}.$$

The first holds with equalities only if $\deg^-(v)=1$ for all nodes in layer i and $\deg^+(v)=k$ for all nodes in layer $i+1$; the second holds with equality only if $\deg^-(v)=l$ for all nodes in layer i and $\deg^+(v)=1$ for all nodes in layer $i+1$. If $\deg^-(v)=l$ for all v in layer i and $\deg^+(v)=k$ in layer $i+1$, then $l r_i = k r_{i+1}$, and the following relation holds

$$r_{i+1} = \frac{l}{k} r_i .$$

Taking into account boundary conditions $r_0 = m$ and $r_{h+1} = n$ gives

$$r_i = \left(\frac{l}{k}\right)^i m = \left(\frac{l}{k}\right)^{h+1-i} n \quad \Longrightarrow \quad \left(\frac{l}{k}\right)^{h+1} = \frac{n}{m} .$$

Equation (3) is obtained by taking the logarithm of the above equation. Equation (4) is obtained by substituting it into $s = \sum_{i=1}^{h} r_i$ for the number of hidden nodes.

Remark 7. Observe that the ratio $l/k = n/m$ gives $h = 0$ and $s = 0$. This suggests inequality $l > kn/m$. On the other hand, $\lim_{l \uparrow k} h = \infty$, so that $l < k$.

Example 8. Substituting numerical estimates for m and n in human CNS (see Appendix), and using $k = 2 \cdot 10^3$, $l = k - 1$ gives $h = 9{,}461$ and $s = 0.96 \times 10^{11}$. Once again, the latter number is close to the estimated number of neurons in a human brain (10^{11}).

4 Discussion

Human nervous system and brain in particular is an example of extreme and ingenious engineering by nature. Apart from problems of optimal perception, communication and control it solves problems of minimization of building material and energy consumption. Assuming that the conflicting objectives correspond to maximization and minimization of connectivity, we derived some simple relations between numerical parameters of a graph representing the brain network. The relations can predict some of the parameters about human nervous system, estimated from biological data.

The models discussed are of course simplifications of real nervous systems. Many properties were not taken into consideration. For example, it is known that motor neurons have some of the highest number of synapses in the brain, while neurons that are close to perceptual organs have some of the smallest number of inputs [1, 4]. Thus, brain networks are not regular graphs. Taking into account variable connectivity is one possibility to refine this work. Another property that was not explored here is path connectedness between sources and sinks.

Models that can be evaluated on data and have the ability to predict natural phenomena are of a particular interest. Understanding the basic principles of organization of the brain can help in optimization of other networks, such as communication, social and distribution networks.

Appendix A Estimated Parameters of Human Nervous System

Human nervous system is arguably the most complex, but at the same time one of the best studied neural network. Human nervous system, as well as of other vertebrates, is organized into the central nervous system (CNS), which consists of the brain and the spinal cord, and the peripheral nervous system (PNS), which consists of the somatic and autonomic nervous systems. PNS is responsible for collecting all the sensory information and sending all the control signals to the body, which include voluntary actions, sympathetic and parasympathetic processes. CNS is insulated from the rest of the body by three layers of tissue, called *meninges*, and it is connected to the environment (body) by nerves, which carry all the fibers between CNS and PNS [4].

The brain is by far the largest collection of neurons in the body with some estimates on the order of 10^{11} neurons, while the spinal cord contains approximately 10^9 neurons, many of which aggregate and relay the information into and from the brain [1]. The brain is likely to fulfill the majority of information processing and control functions in the body. Thus, for a directed graph G representing human CNS, the number of hidden nodes is approximately $s = 10^{11}$ (i.e. estimated number of neurons in the brain and spinal cord).

The numbers m and n of sources and sinks of G are respectively the numbers of *afferent* (input or ascending) and *efferent* (output or descending) fibers of the CNS.

Table 1 Summary of afferent and efferent fibres in cranial nerves. Numbers of fibres found in [1,4] are shown in *roman*. *Italics* show estimates for nerves with no data found, and computed as average numbers based on the other cranial nerves

Nerve	Afferent (IN)	Efferent (OUT)	Fibers
Olfactory	Smell		$1.2 \cdot 10^7$
Optic	Vision		$1.2 \cdot 10^7$
Vestibulocochlear	Hearing, balance		$3.1 \cdot 10^4$
Oculomotor		Eye, pupil size	$3 \cdot 10^4$
Trochlear		Eye	$3 \cdot 10^3$
Abducens		Eye	$3.7 \cdot 10^3$
Hypoglossal		Tongue	$7 \cdot 10^3$
Spinal-accessory		Throat, neck	*$1.1 \cdot 10^4$*
Trigeminal	Face	Chewing	$8.1 \cdot 10^3$
Facial	2/3 taste	Face	10^4
Glossopharyngeal	1/3 taste, blood pressure	Throat, saliva secretion	*$9 \cdot 10^3$*
Vagus	Pain	Heart, lungs, abdominal, throat	*$9 \cdot 10^3$*

It is widely believed that $m \geq n$. Numerical estimates of these parameters for human CNS can be done by adding up data for the numbers of fibers in individual nerves [2]. Here we describe briefly this method and results.

There are 12 pairs of cranial nerves that connect directly to the brain, and 31 pair of spinal nerves that connect to the spinal cord. The majority of the nerves carry both afferent and efferent fibers. Table 1 shows numbers and types of fibers in cranial nerves [1, 4]. Note that we did not find data for spinal-accessory, glossopharyngeal and vagus nerves, and used estimates from other similar cranial nerves. The estimates are shown in italic. Thus, the total numbers of afferent and efferent fibers in cranial nerves were estimated as

$$m_c \approx 2 \cdot (1.2 + 1.2) \cdot 10^7 + 2 \cdot (31 + 4.1 + 5 + 4.5 + 4.5) \cdot 10^3 = 4.81 \cdot 10^7$$

$$n_c \approx 2 \cdot (30 + 3 + 3.7 + 7 + 10.9 + 4.1 + 5 + 4.5 + 4.5) \cdot 10^3 = 1.45 \cdot 10^5.$$

Spinal nerves are both sensory and motor, so that each spinal nerve carries both afferent and efferent fibers. The numbers of these fibers had to be estimated from cranial nerves due to lack of data. These estimates for the total numbers of afferent and efferent fibers in spinal nerves are

$$m_s = n_s \approx 2 \cdot 31 \cdot 4.5 \cdot 10^3 = 2.8 \cdot 10^5.$$

Adding together our estimates for cranial and spinal nerves gives the following numbers of parameters m (number of sources or afferent fibers) and n (number of sinks or efferent fibers) of human CNS:

$$m = m_c + m_s \approx 4.84 \cdot 10^7$$
$$n = n_c + n_s \approx 4.26 \cdot 10^5.$$

Finally, the number of synapses of an average neuron is estimated to be in the range $10^3 - 10^4$ [1, 4]. These numbers allows us to define constraints on the in-degree $\deg^+ v \leq k$ of hidden nodes and sinks.

Our estimates, although not very precise, enable us not only to appreciate the incredible complexity of human CNS, but also provide some qualitative information. In particular, they support the inequality $m \geq n \geq \deg^+(v)$. The estimates can be used to evaluate our hypothesis about maximal connectivity of brain networks with constraints.

Acknowledgements This work was supported in part by EPSRC grants EP/DO59720 and EP/H031936/1.

References

1. M.F. Bear, B.W. Connors, M. Paradiso, *Neuroscience: Exploring the Brain*, 3rd edn. (Lippincott Williams & Wilkins, PA, 2007)
2. R.V. Belavkin, *Do Neural Models Scale Up to a Human Brain?* International Joint Conference on Neural Networks (IJCNN 2007) (IEEE, NY, 2007)
3. E. Bullmore, O. Sporns, Complex brain networks: Graph theoretical analysis of structural and functional systems. Nat. Rev. Neurosci. **10**, 186–198 (2009)
4. R. Poritsky, *Neuroanatomy: A Functional Atlas of Parts and Pathways* (Hanley & Belfus Inc., PA, 1992)
5. C.E. Shannon, A mathematical theory of communication. Bell Syst. Tech. J. **27**, 379–423, 623–656 (1948)
6. R.L. Stratonovich, *Information Theory* (Sovetskoe Radio, Moscow, 1975), In Russian